Reading and the Making of Time in the Eighteenth Century

Reading and the Making of Time in the Eighteenth Century

CHRISTINA LUPTON

Johns Hopkins University Press
Baltimore

Printed in the United States of America on acid-free paper
2 4 6 8 9 7 5 3 1

Johns Hopkins University Press
2715 North Charles Street
Baltimore, Maryland 21218-4363
www.press.jhu.edu

Library of Congress Cataloging-in-Publication Data
Names: Lupton, Christina, author.
Title: Reading and the making of time in the eighteenth century / Christina Lupton.
Description: Baltimore : Johns Hopkins University Press, 2018. | Includes bibliographical references and index.
Identifiers: LCCN 2017044508 | ISBN 9781421425764 (hardcover : alk. paper) | ISBN 1421425769 (hardcover : alk. paper) | ISBN 9781421425771 (electronic) | ISBN 1421425777 (electronic)
Subjects: LCSH: Books and reading—History—18th century. | Books and reading—Social aspects—History. | Book industries and trade—Europe—History—18th century. | Time—Psychological aspects—History.
Classification: LCC Z1003 .L928 2018 | DDC 028/.909033—dc23
LC record available at https://lccn.loc.gov/2017044508

A catalog record for this book is available from the British Library.

Special discounts are available for bulk purchases of this book. For more information, please contact Special Sales at 410-516-6936 or specialsales@press.jhu.edu.

Johns Hopkins University Press uses environmentally friendly book materials, including recycled text paper that is composed of at least 30 percent post-consumer waste, whenever possible.

To all readers without time to read

Contents

Acknowledgments

This book owes much to the last set of people to read it in manuscript form: Ian Balfour, Christina Britzolakis, Maria Damkjær, Lynn Festa (not just for culling the colons), Marcie Frank, Matthew Garrett, Heiko Henkel, Michael Meeuwis, Mark Philp, Adela Pinch, David Taylor, and Chantal Wright. Not only did they fix mistakes and contribute significantly to this book's final version, their wisdom and generosity kept me company through a long Danish winter of rewriting.

Just as materially, I owe much to the Alexander von Humboldt Trust and to the Leverhulme Trust. As I say in greater length in the conclusion, I have written with the gifts of time their grants gave me in mind. I am thankful to Matt McAdam for taking this project on in an early incarnation, for the team at Johns Hopkins University Press for bringing it to life, and to David Goehring for his copyediting. And I am grateful to funding from the Warwick Humanities Research Fund, which facilitated conversations vital to my thinking over the past few years.

An important part of this book came together in a paper at the English Institute, and I'll always be glad Sandra Macpherson, Sangeeta Ray, Bill Brown, Bradin Cormack, Jackie Goldsby, Martin Harries, Jonathan Kramnick, David Kastan, Doug Mao, Meredith McGill, and Frances Ferguson invited me into that conversation and grateful for the work they and many others do in keeping the EI alive as a space of thought and conversation.

I continue to learn so much from workshops and conversations around the world. Thank you to Leslie Adelson, Emily Hodgson Anderson, Samuel Baker, Liz Barry, Christian Benne, Paddy Bullard, Miranda Burgess, Tita Chico, Mark Currie, Nan Da, Alex Dick, Dennis Duncan, Anne-Lise François, Kristen Girten, Danny Hack, Sarah Haggarty, Sarah Kareem, Jeff Knight, Marjorie Levinson, Kathy Lubey, Deidre Lynch, Sarah Mesle, Tom Mole, Sarah Moss, Jesse Molesworth, Brad Pasanek, David Porter, Leah Price, James Raven, Helen Small, Andrew Piper, Jonathan Sachs, Elisa Tamarkin, Gill Partington, Bruce Robbins, Emily Rohrbach, Sean Silver, Rebecca Spang, Kirsten Tranter, Chad Wellmon,

Gillian White, and Eugenia Zurokski for your invitations, dialogues, and arguments. All of you are a constant reminder that having time to read is no good without time to talk, and that writing makes no sense at all without good readers.

In writing about eighteenth-century women like Talbot and Carter, Opie and Inchbald, I have stood squarely on the shoulders of scholars including Betty Schellenberg, Judith Hawley, Ros Ballaster, Karen O'Brien, Elizabeth Eger, Jennie Batchelor, and Terry Castle. Their work has allowed me to write about women readers and writers without this having to be a book about gender or a history book as such. My debt to their scholarship is greater than my endnotes suggest.

I have also relied heavily on work done by historians, especially James Raven, Jan Fergus, Carolyn Steedman, and Jonathan Rose—again, the fact that this book is not a proper history of reading was only possible because there are so many good ones out there.

I thank the editors of *ELH* for their permission to republish work that first appeared there. Parts of chapter 1 were published as "Immersing the Network in Time: From the Where to the When of Print Reading," in *ELH*, 83.2 (Johns Hopkins University Press, 2016), 299–317. Parts of chapter 3 appeared as "Contingency, Codex, the Eighteenth-Century Novel," in *ELH*, 81.4 (Johns Hopkins University Press, 2014), 1173–92.

Martin Ryle, Kate Soper, Zara Lupton, and Chris Davies all continue to model the good life in different and inspiring ways. Camilla, Tina, Jesper, Sara, Kristian, Ludvig, Sofus, Eja, Johann, and Villads make and give time every week in the form of the Spiseklub, and I will always think of this book in connection with their friendship and food. Heiko read it all from cover to cover and did a lot of other things that made writing it possible and that make the future beyond it seem bright. Maibritt and Perry were and are just their lovely selves, good to live and think with, story makers and gift givers of the very best kind.

Reading and the Making of Time in the Eighteenth Century

When Do We Read?

There is a Moment in each Day that Satan cannot find
Nor can his Watch Fiends find it, but the Industrious find
This Moment & it multiply, & when it once is found
It Renovates every Moment of the Day if rightly placed

—Blake, *Milton* 35:42–45

THE SHORTNESS OF TIME

This is a *when* book: When do we read? When do we read *books*, I mean, because I know when we read all the rest—the emails, the texts, the news, the status updates, the articles posted by friends, the lists we make of all the things we want to read. We read them all the time, in every crevice between other activities when our hands, or eyes, are free. We read them when our children are in the bath; when our friends are waiting to talk to us; in meetings when our colleagues aren't looking; when we lie in bed at night, the lights already out but the phone at our side; when we are waiting at a traffic light; or when we are taking a break. Our days are flooded with small texts that contract and expand to fit the minutes we choose to give them. But when do we read books, those texts whose heft seems more plaintively than ever an appeal to reading's duration?

In the years I've been working on this project, there has been a growing sense that reading for any length of time, and with any depth of concentration, is getting harder.[1] There is actually very little evidence that the purchase or production of paper books is in decline, and plenty of cases are being made for the viability of paper as a medium that might continue to coexist happily with screens. Reading itself, understood in terms of word count and frequency as an event, is certainly on the rise.[2] But the claims for books' neglect stack up: we once had more time, more servants, proper summer holidays, fewer distractions, longer childhoods, better concentration. Students read more then, getting lost in books, whereas students and children today can barely make it through a chapter without seeking relief from other media. We used to work fewer hours, leave the office, read at night—not have phones. Before word searches and big data made reading computational, we read books. Leisurely, inquisitive, slow reading gets attached imaginatively in all these ways to that idea of a past in which we were more effectively saturated by books.

My purpose in looking back to that first era of widespread book reading is not, however, to emphasize the abundance of deep reading, or book reading, or leisurely time that we have lost. The more I've studied readers of the eighteenth century, the more I've doubted that we (by which I mean a historically fairly new "we"—people who can buy books but also must earn money, manage households, walk dogs, bathe children) ever really had more time to read. I do not believe that the minutes crowded by messages, HBO series, and childcare today correspond in any direct way to time that we—posters and messengers, scavengers of the internet, wage workers and intellectuals—once spent with books. The readers I represent struggled to make room for the reading of books in lives that they perceived, rightly or wrongly, to be crowded in the same way we perceive ours to be. They worry, like us, about other media forms that seem quicker and shallower and more enticing than books. They sense that round-the-clock entertainment and distraction might render book reading extinct. They dream of a future when books will find a wider and more attentive public. In focusing on book reading rather than on media consumption generally, my first gambit, then, is this: ever since people like us have had access to books, the time we've spent with them has been defined as fragile, hard to come by, and good to hope for.

Time in this project does not represent a vector along which late modernity picks up speed or the horizon of mortality toward which Heidegger orients us. I am only tangentially interested in time as something that can be deep or shallow, slow or fast. More important here is my sense of time as an ongoing axis of struggle and possibility: one along which our relationships to each other, to the world, and to objects and our labor are arranged and spaced out and joined up, often with elasticity and creativity. I am interested in the idea of education conceived of as a giving of time, and I am interested, in a fairly quiet way, in gender as it plays into that project. I am on the side here—although many caveats will ensue—of David Harvey, for whom "temporal practices are never neutral in social affairs. They always express some kind of class or other social content, and are more often than not the focus of intense social struggle."[3] Some of my cues come, too, from writers like José Esteban Muñoz, for whom occupying time differently is a political strategy instantiated through but relevant beyond the site of cultural production.[4] And while my focus is on media—in this case, on bound pages, paper, and print—as agentive in making time apparent to us in new ways, I have an eye to those arguments for technology as a platform for resisting the feeling of time's shortage and speed. Judy Wajcman, for instance, advocates human control of the mechanisms that seem to accelerate and steal time today: Why not the car that makes us go slower? Or the e-book that shuts out worldly distraction?[5] Mark Hansen, working in a very different register, promotes digital media as producing new kinds of temporality, both along with and aside from human cognition, and points to the sites of radical creativity and new views of the future emerging with technology.[6] These arguments focus on twenty-first-century developments, but I want to wrest them back, to pull them out of shape to see what they might teach us about why reading of books in particular has been, and continues to be, a juncture where technical and human agents collaborate fiercely in creating much-desired and nonlinear experiences of time.

The eighteenth-century readers I describe include actors, clergy, professional novelists, translators, housekeepers, and politicians. Most of them struggle, in terms that should resonate in the twenty-first century, with the fear that time for the kind of reading they associate with books (rather than newspapers or sermons or letters) is always too short. We find them complaining in their correspondence, diaries, notebooks, and published writings about not having time alone, about the pressures of work and sociality, and about lives of diversion that make meaning and purpose difficult to grasp and productivity hard to judge. With only a few exceptions, these are readers who anticipate our "we" by being people who also have casual access to many kinds of text as well as some form of work that they must do, some other kind of labor that competes for their time: they are the professionals, not the aristocrats, of their age. Often they are also readers who believe, more typically than we might think, that the reading of books is an occupation endangered by newer forms of entertainment and print consumption. Against these odds, they are in evidence as good scholars—makers of systems, tricks, patterns, and revolutions that allow them to develop as readers despite the odds. They deploy books as the time-turners as well as the time-tellers of their modern lives.

As social history, this argument involves not so much a new tale as one that joins up in new ways familiar narratives about the eighteenth century. It has been well established that the spread of clocks and calendars and institutions in this period introduced a new kind of chronometry to everyday lives. This is the era in which Walter Benjamin's homogenous empty time becomes thinkable as an unending and regular sequence of minutes against which any other way of being untimely must come into relief. It is also when history ceases to be experienced as predominantly cyclical and becomes visible instead as a force propelling us into what Reinhart Koselleck describes as the "open future." Since the second half of the eighteenth century, he argues, "history no longer takes place in time, but through time. Time is metaphorically dynamicized into a force of history itself."[7] These newly regularized frameworks of daily and historical time provide the matrices into which various personal rhythms and projects of national and institutional life begin to fit—and against which others emerge.

It is also over the course of the later 1700s, Foucault observes, that discipline ceases to be a matter of corporal punishment and becomes a matter of controlling the prisoner or subject in time.[8] And the same years are those that E. P. Thompson associates with the emergence of "time-discipline," a convergence of Protestant and nascent capitalist forces that encourages a clocking of labor as alienated hours spent at work. Between 1780 and 1830, he argues, "the 'average' English working man became more disciplined, more subject to the productive tempo of 'the clock.'"[9] Thompson makes this argument succinctly in "Time, Work-Discipline, and Industrial Capitalism," where he suggests that it is "by the division of labour, the supervision of labour; fines; bells and clocks; money incentives; preachings and schoolings; the suppression of fairs and sports—new labour habits were formed, and a new time-discipline was imposed."[10] In Thompson's account, the perception of the worker's time as something to be used profitably erodes other possibilities, both past and future, of less instrumental ways of being in the world.

Yet we also know that the availability of print material increased phenomenally in this period. Not only did the quantity and diversity of books grow, so did their accessibility and the forms in which they might appeal to a time-strapped worker. The price of books in this period fell (though not as radically as it did later), and novels emerged as a new and rapidly expanding segment of the print market. To be sure, making this kind of claim risks subscribing to the too-much-to-know thesis, according to which almost any group of readers in history can be presented as having struggled in ways analogous to our own with unprecedented new quantities of reading material.[11] But the work of describing and quantifying this rise for the second part of the eighteenth century has been done meticulously by James Raven, who has elaborated throughout his career on the generally accepted thesis that middle- and working-class readers began to consume fiction during these years. In multiple studies, including *The Business of Books: Booksellers and the English Book Trade 1450–1850* and *Bookscape*, Raven shows how the growth of printing shaped everyday life for the English reader.[12] These are years in which newspapers, magazines, and print merchandise rained down upon the eighteenth-century consumer at new speeds, but also years in which books, figured in antiquarian collections, private and circulating libraries, and streetscapes dotted with new printing houses and bookshops, piled up.[13]

But how exactly do these two phenomena add up? How, with less unstructured time available than ever before, and with time equated increasingly with money, could the eighteenth-century worker commit so many hours to reading that wasn't directly productive? If it is really the case that this was the age in which the private purchase, consumption, and circulation of books picked up speed exponentially as an activity, how are we to square this with the fact that the hours in which normal people were at ease were dwindling at an inverse rate? What about women, who were kept busy in so many ways that worked against them becoming readers: In what hours did they become the most infamous of the novel's new consumers? Historians of reading working in this period have attended carefully to the question of how people accessed books by using publication, purchasing, and library records as well as book clubs, reprinting, and the surge in fiction reading as evidence of the book's new place in working lives.[14] Jonathan Rose, focusing on later centuries, draws richly in *The Intellectual Life of the British Working Classes* on readers' own accounts of how they accessed books, of where they read them, and of what they made of them.[15] But isn't the more pressing question how these people found the time to read at all? As Robert Altick comments in *The English Common Reader*, "Obviously one cannot read without some leisure in which to do so."[16] The paradox seems to be of the most basic kind: if work time, clock time, and instrumental time use become signatures of life for common people in the eighteenth century, most of whom still work unthinkably long hours, how are we to explain the hours they sat with books in this period? How did the availability and length of these newly available texts, perceived both as things that needed and rewarded time spent with them *and* as dangerous diversions from spiritual and economic labor, contribute to the period's economy of hours?

It is possible to argue, of course, that print was a vehicle for bringing people into modern time and that the content of novels, news, sermons, and also their media, helped standardize and historicize new ways of being. This can be observed of the many texts that became everyday instruments of time-keeping: conduct books for days of the week, calendars and almanacs with blanks to be filled in, collections arranged as annual rosters of reading. It's most evidently true of newspapers and journals, punctual forms of print publication that, as Benedict Anderson has famously argued, enlisted readers as a national body, constituted because people understood themselves to be reading the same news at more or less the same time.[17] These punctual forms of print publication have also been closely connected to the positive kinds of reading and writing that allow a reflexive public sphere to emerge in the eighteenth century. Jürgen Habermas's emphasis falls on the physical and institutional spaces that sponsor a well-functioning public sphere, with his examples ranging from eighteenth-century British coffee houses where news was shared to the Muslim radio stations and mosques that would be necessary in his view in a properly liberal Germany today.[18] Michael Warner's *Publics and Counterpublics* stresses time much more as the dimension in which readership emerges, with daily and weekly rhythms of publication being vital in his view to the creation of modern American publics. Writing in 2002, Warner speculates about whether internet publication, which lacks that punctual form, can have the same effect as regularly published newspapers and journals in making readers aware of themselves as part of a public.[19]

But it is against the background of newspapers, periodicals, almanacs, and sermons appearing regularly that books and their consumers gain their own special relation to time in the eighteenth century. Even as the length of time between printing and end-sale publication diminishes generally in this period, book reading develops its own character as an activity valued because it can offset newer and faster kinds of reading. Books pile up, get given, preserved, recycled, purchased, deferred, and absorbed at special rates because they are not punctual or as urgently demanding of attention as news or occasional writing. They get anxiously and sentimentally defended as old-fashioned. They begin to get read, as Deidre Lynch shows so well in *Loving Literature: A Cultural History*, in newly affectionate cycles of return:

> It is tempting to imagine that if we could time-travel and take with us an especially sensitive stethoscope, we might as visitors to Britain around the year 1830 actually be able to hear, as if it were a heartbeat, or a kind of bass line, pounding beneath the louder noise of public history, the rhythm that the inhabitants steadily beat out as, turning pages they had turned before, often at the same time of the week or year as before, reciting according to schedule the familiar words they had recited before, they conformed to their bookish routines.[20]

Less devoted readers than Lynch's also find more syncopated beats to read to, individual and collective rhythms of opening and sharing books, determined, for instance, by weather, holidays, annual book fairs, library opening times, or lulls in theater programming.

To be sure, the codex format with its continuous, bound pages can also be connected to the invention of clock time and the insertion of the reader into it; to eighteenth-century economies of time, rather than their corruption. The most direct argument for the way eighteenth-century books harness readers to a larger temporal economy is Stuart Sherman's erudite *Telling Time*, which describes the diary and the periodical helping users insert themselves into time as a successive series of minutes: "A new construction of time as series within series, concentric and cumulative, beginning with the small intervals clicked out at the clock's core, and radiating outward to the markings on the dial, to encompass a whole system of measurement and calibration: ticks, seconds, minutes, hours and (on calendrical clocks) days and years as well."[21] In this framework, pages represent hours and days, their structure inviting a diary-keeper like Samuel Pepys to coordinate his life on paper. Sherman sees these closely intertwined technologies of life-writing and time-keeping represented in the first generation of British novels, whose diurnal form leaves more occasional and intermittent forms of narrative behind. The pages of the book give materiality to the tick, tick, tick of modern life, with texts like *Robinson Crusoe* and *Pamela* indebted to this environment for their own representation of one minute following the next. However belated, irregular, or circular reading may be, Sherman suggests, these fictions train readers to imagine even the most personal life narrative occurring in regular, daily time.

More media-centered accounts of print in this period have also made it possible to think of codex as a technology that co-operates at the point of its reception with a modern temporal economy. As packages of information, Chad Wellmon and Brad Pasanek have argued, books have long supported the idea of efficient access.[22] This story can be told in material terms by looking at the abridgments, reviews, indexes, catalogs, and abstracts that quickly appeared alongside books as ways to speed up their consumption. Searching, one of the forms of literacy that seems most directly to threaten sustained book reading now, has long been associated with the book.[23] Even as books were taking up more space on people's shelves as signs of leisure and learning, they were developing as technologies that allowed text to be plumbed quickly. Wellmon, looking at the late-eighteenth-century German case of "information overload," describes indexes, charts, and encyclopedias as explicitly designed to contain and synthesize this excess.[24] His focus is on a broad form of cultural endeavor, an institutional response to print production that leads finally to new technologies of disciplinary inquiry and subject formation channeling that surplus of book-bound knowledge. He is less interested than Thompson in the class implications of this reorganization. But there is a link between his argument and the context Thompson describes: the feeling of there being too much to read relates closely to the problem of people losing the leisure that would support unstructured, literary, or general reading—and, more precisely, of a new kind of consumer emerging for the book and its institutions for whom that time was never available. Those technical fixes to the problem address this particular reader's lack of time as part of the era's surplus of print knowledge. They suggest that the new kinds of text searching we do now extend what has always been the capacity of the book to be read quickly in a way that differentiates it from scrolling technologies that are much less easily indexed or sped up.

I find these arguments for the book as a technology that co-operates in Thompson's "time-discipline" largely convincing. With this project, however, my focus is on the *reading* of books as an activity more recalcitrant and resistant to efficiency than their indexicality and disciplinary identity suggests. In the stories I tell, people pick up books, reread them, and postpone reading them in ways that are often out of kilter with the idea of modernity's commitment to regularity and speed. Reading shows up as an activity that involves irregular, stolen, and anticipated moments as often as it does routine or synchronized or profitable ones. If print production belongs to a history of discipline and efficiency, then reading, especially of books, also belongs to a different history that has as much to do with irregularity and the dream of revolt against those regimens of productivity.

There's a personal dimension to this. The year in which I thought most intensely about time, and about the arguments I lay out here, was one in which I worked long hours as a university administrator. I rushed around campus from meetings that were so tightly scheduled I could barely make it from one side of campus to the next. I spent my evenings reading emails and making spreadsheets and writing reports. I caught trains at dawn and often got back to London late at night or just in time to pick up a child; I organized conferences that left me no time to prepare except at airports. "I have no time," I thought, "no time at all." And yet it was at that very ebb of intellectual life, that very point where my days felt more scheduled and more tightly packed than they ever had before, that I began to think about what reading books was to me. I became full of desire for quiet settings—hammocks, sofas, farmhouses, empty rooms, long evenings—in which I would imagine myself with a book. I'd go sometimes at the end of a day to visit a friend who was sick and waiting to hear news. I'd arrive late, breathless, but bearing books as gifts, and feel the very faraway prospect of reading, sensing how different our worlds were. Over Christmas and on trains and planes I'd burrow into novels, autobiographies, old theory books. I'd read with my kids bouncing around my shoulders. I wrote almost nothing: there seemed no time to be won there, and though I snuck brief glances at news and blogs, it was my hunger for longer reading that grew fierce. It's partly out of that paradox that I come now, in days much quieter and more privileged, to write about eighteenth-century readers for whom book reading in particular was much less obviously aligned to modern time and its pressures than the arguments of Anderson or Sherman or even Wellmon might suggest. My interest is in a literary and historical understanding of book reading as something that has been used to cut and complicate homogenous empty time—but also to remake temporal experience creatively in ways from which digital readers, modern workers, and those thinking today about education might still learn.

THE TENSE OF READING

One text, *Low-Life: Or, One Half of The World Knows Not How the Other Half Live*, published anonymously, probably in 1750, has been with me all along. This text participates in the regulating of time that Thompson and Sherman describe. Its hundred-odd pages are divided into twenty-four chapters, each of which scans the activity of a range of Londoners at one hour of the day. At one o'clock, for instance, we hear of "Taylors, whose work is in a hurry, leaving off Labour, in order to take two or three Hours Sleep, that they may be able to make fresh Attack on Business."[25] At six in the morning, we glimpse, among others, "beggars, who have put on their woeful countenances, and also managed their Sores and Ulcers so as to move compassion . . . carrying Whads of straw to the Corners of the most Public Streets" (32). And before noon, we see "Poor People that lodge in Low Rented houses, going to each other, and after paying their aukward compliments, borrowing saucepans and stewpans" (55). At six on Sunday evening "Authors of both Prose and Verse, whose Wives and Childrens clamorous Tongues prevent their studying in their own Apartments, are slowly walking about remote Parts of the Town, with Memorandum Books in their Hands, and taking down Notes of their best Thoughts, that they may digest them into proper Order, when they can be quiet in their own Lodgings" (80).

This survey continues to clock the activities of rich and poor, honest and corrupt, male and female, throughout the day and across the city. The author compares his project to Hogarth's graphic one, describing himself aiming "to deliver the Actions of every hour, as they really pass; omitting nothing, however trifling it may seem, which is the subject of that Hours employment or Abuse" (v).

The first thing to note here is that *Low-Life* shows a diverse community of people connected not in space but in time. Every page turned aligns the city's population with the calendar as well as the clock. It is June 21, the start of summer, and the hours we read about begin at midnight on a Saturday and end at midnight on a Sunday. In some ways this can be read as Anderson's nation. The whole premise of *Low-Life*, as its subtitle suggests, is that even people who share space can be ignorant of each other. Yet homogenous time, it begins to seem, is the matrix everyone is plugged into. Lloyd Pratt describes such temporal unity in this way:

> The world is ruled by the passage of only homogenous-continuous/self-contemporaneous time: this kind of time pre-exists and preforms the world; everyone everywhere has the same time in common. Consequently, this notion of time suggests that one can represent the totality of a given moment via a cross-section, as all the constituent components of a totality are both defined and restricted by their shared participation in the same interval of time. No one can get ahead of or fall behind time, because homogenous-continuous/self-contemporaneous time, as the medium of our existence, restricts and subordinates the world to its even and linear folding. This time resembles a series of regularly spaced paving stones rising up to meet the world and, in rising to meet it, allowing the world to unfold.[26]

This "paving stone" theory of time seems assumed by *Low-Life* as a text that represents all the habits and institutions of time-keeping and is itself structured as a succession of hours. And *Low-Life* also confirms that the activity most connected with this view of time, as Thompson's argument might lead us to anticipate, is work. If authors, nurses, and prostitutes become visible in interesting ways as the inhabitants of London's spaces, they come into focus in *Low-Life* as workers whose long hours of labor can be counted. While its cross-sections of city life show people reveling, drinking, sleeping, and walking, *Low-Life* stands out for its representation of people working for a living around the clock and throughout the week.

However, there are certain distortions of clock time introduced by the structure of *Low-Life* that anticipate arguments such as the one Pratt goes on to make in *Archives of American Time*, for the "homogeneous-continuous" version of time having come under pressure since its early days. For a start, *Low-Life* is not set in normal time. It is set on a Sunday, a day many cherished as a break from normal time. An eighteenth-century Sunday, we'll see in the next chapter, gave a welcome time frame to reading, an activity that was in turn condoned by many as a way to help preserve the distinctiveness of the Sabbath. But reading is fairly low-key as an activity represented in *Low-Life*. There are "publick Prints at Coffee-Houses" (58) and "the News-Papers in every little Ale-House" (90), but the actual reading that goes on is limited and superficial: "Plays and Romances" are read by the ladies about St. James (54) and Bibles are shuffled through in church (54). We see "Physicians in their Chariots, poring over Books, like Malefactors going to the gallows, to give the Town a sense of their Religion, or rather deep study, when perhaps, what they are reading may be a ludicrous Pamphlet or political News-Paper published the Day before, and filled with bitter invectives against the Government" (45). Servant boys are shown cramming a few printed sermons in an attempt to show they've been at church when they haven't (50). Other than that, not as much seems to be read as is written (several authors appear alongside the harassed father-poet, all hack writers and workers of a kind). There are certainly none of the sociological descriptions we get later, for instance in Richard Hoggart's *Uses of Literacy* or Janice Radway's *Reading the Romance*, of workers intent on reading, hunkering down in busy rooms, finding room at a table, and screening out noise and distraction by disappearing into a text.[27]

Yet the question of reading is posed in other potentially more significant and disruptive ways by *Low-Life*. Its own reader, for instance, is positioned by the structure of the text as someone who must pass time with its characters. Because London is represented in slices more geographically and socially disparate than anyone could join up at a glance, even a long one, the connection with Hogarth's painting is soon dropped. The text becomes caught up in its own representation of time in a way that visual art is not. As we turn pages, it becomes clear that reading about work requires an outside, a fold in work-time, that just looking doesn't. It requires the very Sunday from which it seems precluded here. *Low-Life* skirts the question of people's Sunday reading, and yet any book, even a short one like this, that accounts so diligently for the minutes of the day, must bring up the question of its own consumption. When was *Low-Life* to be read?

One possibility is that it is written to be read in hours reclaimed from the brutal kinds of round-the-clock entertainment and labour it describes: "If by representing these serious Trifles," the author writes in the preface, "I could but persuade my Readers from pursuing them, and engage them in Nobler aims, the end of this Work would be answered" (v). This would position *Low-Life* at a strange vanishing point, suggesting the Sunday reporter in the ideal world would have nothing to describe but the fact of everyone reading *Low-Life*. But the tone of the booklet belies this suggestion. The author's disapproval of the activities described isn't severe enough to make their elimination seem genuinely desirable, and the text itself is not pious enough to replace them as proper Christian entertainment. As a text flagging the fact that it must be read sometime, *Low-Life* invites another calculation. While toying with synchronizing reading and the clock (each hour corresponds to a four- to five-page chapter), the time it takes to read a chapter of *Low-Life* is much less than an hour. *Low-Life's* reader moves palpably faster than the clock, and with special freedoms given by the codex to skip ahead or stop time's progress.

This becomes interesting because of the strange tense in which *Low-Life* is written. Carolyn Steedman identifies it as dominated by free existential clauses without tense: "Young ladies," for example, are described "demanding adoration instead of paying it in churches." "The effect of *Low-Life* is produced," Steedman argues of such description, "by the absence of tenses, and thus of time. What happens in its pages happens outside calendar and clock time (though it is presented as entirely framed by those chronological, linear, and real measures of time)."[28] Her conclusion is that this absence of time makes *Low-Life* a text without referent. In historical terms it is not, she argues, about anything: it has no topic other than itself. As statements, "young ladies demanding adoration" or "kittens in the grass" do not refer to history: they do not need young ladies or kittens to have existed or to have done anything at any particular time in order for them to be true.

Free existential clauses are, for instance, also the time of stage directions: "x enters the stage." That event is real (it does not belong to the fictional time of the play), but it is almost impossible to map in terms of homogenous-continuous time. As Steedman's analysis suggests, this creates an incommensurability between the time of reading, which is never then, or now, or in the definite past, and the time of historical events. Reading takes time, but it does not compete directly with things that have happened, or are happening, or will happen. The process of workers reading seems instead to be ongoing, like the text itself, and therefore difficult to place in a historical or even daily continuum. *Low-Life's* reader did not read then, does not read now, and will not necessarily read in hours to come. But somehow, its reader reads. Books are getting read. *Low-Life* imagines this by grammatically locating the reading of events situated within diurnal time outside that framework. The tense of the book's reading refutes the time-discipline to which the text makes all human activity answerable, becoming its own solution to the shortage of hours for work, worship, and leisure that its portrait of time use demonstrates. The most convincing conclusion to be drawn from *Low-Life* is that reading about life is not directly in competition with living it.

LITERATURE AS RESISTANCE

Here then is an answer. When do we read? We read in grammatically improbable tenses not easily accommodated by descriptions of time. Whatever we read, I'll argue in the chapters to come, we read in the interstices of time that grammar can help flag, in the future anterior, in time opened up by contingency as an awareness of what could have been otherwise, and in the time that has not yet come. These temporalities refer to the time in which books are to be read rather than to any conventional ones in which their narratives take place. Although there are many reasons that certain kinds of texts (fiction set in the future, poetry) seem to kick back against the measurement of time in special ways, I want to stick here with the idea that book reading's interference with time is not primarily about genre. This approach modifies more familiar claims for the way literature intersects, resists, or co-operates through its content with the temporal economies imposed by modernity. But it also challenges how we define literature in the first place.

We can return here to Thompson's gesture in "Time, Work-Discipline, and Industrial Capitalism," which reserves as its conclusion a view of the special role literary texts play in the late-eighteenth and nineteenth centuries. Thompson doesn't discuss directly the relationship, negative or otherwise, between the historical moment at which the worker's life becomes measured and measurable, and the increased likelihood of this worker having access to large quantities of reading material. But he does gesture to his own twentieth-century students (the article appeared in 1967) as the torchbearers of an older form of life where seasons of excess and leisure mixed powerfully with those of work. A seasonal worker, he suggests, is much like a student swept up in waves of reading before exams but happy to sleep other days away in their entirety. Thompson makes Dickens and Wordsworth his particular allies in connecting this ideal of a pre-modern, seasonal life with a certain modern version of free state education, and of reading in that context as something that would be resistant to economic and temporal calculability and to the binary division of work and leisure. Wordsworth, he argues, presents his own life in *The Prelude* as a defiant response to the demand that hours with books be calculated instrumentally (97). It's reading poetry that gets you out of time.

An equally radical, if less explicitly literary, account of reading's opposition to the logic of capitalist accumulation is offered by Michel de Certeau in *The Practice of Everyday Life*. His affirmative description of the reader (broadly understood here as viewer or listener as well) as "poacher" foregrounds her freedom to ignore a text's linear or progressive logic. His readers appropriate what they need by combining texts as they need them: "Far from being writers—founders of their own place, heirs of the peasants of earlier ages now working on the soil of languages, diggers of wells and builders of houses—readers are travellers; they move across fields they did not write, despoiling the wealth of Egypt to enjoy it themselves."[29] Readers in this construction are immune to the logic of accumulation and expropriation, for reading overlays as free movement even the very texts that represent the logic of historical or capitalist time.

De Certeau's argument seems more polemical than ever these days, when the browsing of web pages and television channels has made this activity less obviously revolutionary, and when the logic of movement can be so easily associated with a particular kind of class privilege. Matthew Garrett queries De Certeau's approach in his study of the picaro, a figure revealing in his terms of the tendency to connect reading with geographical, and then social, movement. Books that appeal to their reader's desire for this trajectory betray for Garrett our incapacity to recognize reading except in these bourgeois terms. Just as we fail in Gayatri Spivak's classic argument to see the subaltern except in terms of the speech that disqualifies the subject from belonging to that category, we fail to recognize anti-bourgeois reading except in terms of the movement that puts the reader on the path to becoming a bourgeois subject. Garrett's argument is not centrally concerned with time, but it surfaces when he suggests that the modern picaro, the migrant laborer, forces us to think about the "when" of reading. He looks to Benjamin Franklin accounting in his *Autobiography* for his extraordinary feats of reading, presented as happening while his fellow workers in London are drinking. And then to J. M. Coetzee's *Youth*, where recent migrant workers in London have physical freedom but are bereft of time: "Coetzee's account of living and working—and reading intensively—in London is, like Franklin's, organized according to the logic of before and after work. But in the case of *Youth*, the drudgery refuses to yield even for the protagonist. At the end of the book—though not, as we know, at the end of Coetzee's story—the central figure is a failure. Swept away from his art by the need to work, Coetzee is preserved only by that before and after time that marks him off from the West Indian workers of Paddington."[30]

Reading in this account resurfaces in the margins of what is represented, appearing as elusive at precisely that point where modern regimens of work close in. Texts like *Low-Life* and *Youth*, having nothing to say about liberation from modern time discipline, pitch their own existence as the impossible possibility of reading that can't be clocked by the text (or by grammar, to go back to Steedman's point about *Low-Life*). When the modern picaro's story leaves no legitimate time for its own telling, the leisure required by reading is forced into view by the novel, not through representation, but as lack. In other words, even these critical accounts of the violence of modern life leave room for reading as the thing for which we stay up late and for texts as instruments that can be played to rhythms other than that for which they are paced, and for narratives as things that turn even relentless workers into readers to whom books might one day, in some other tense, appeal.

THE DIFFERENCE TIME MAKES

I do not want to subscribe too easily to the media-blind terms of de Certeau, for whom the freedom to roam across a page is easily equated with the freedom to change channels or to consume a television series, a season at a time, late at night.

Whatever freedoms come with surfing, browsing, and viewing, they are not, I think, simply equivalent to those related to the reading of bound and printed pages, an activity to which I want to give weight and historical particularity. Nor do I want to end up arguing, as Thompson suggests, that the literary text is differently equipped by virtue of its special content to reverse the temporal logic of capitalism. Or rather, if I believe this in some form, it is less because I believe this of literary genres, and more because I want to bring literature into focus as describing a particular way of holding and using books in time: as a way, for instance, of reading on Sunday or across a lifetime rather than as the generic property of a certain kind of novel or poem.

In making these adjustments, I'll lean on Niklas Luhmann and Bruno Latour, theorists rarely placed alongside critics like Thompson and de Certeau. I initially set out on this project to think about Luhmann in relation to book use. And although my interest in systems theory has waxed and waned in the last five years and continues to raise the hackles of my humanist friends, I've stayed in some sense true to the original ambition of the project, which was to explore the limits of reading as an activity whose effects percolate unevenly through the different spheres of our modern lives, particularly when the temporal dimension of the difference between them is in focus. What is it about modern life, I wanted to know, that makes it so easy to read a book intensely one summer day and ignore having read it at other times, to read it in such different ways at different points in one's life, or to own it and not read it at all? These questions put a slightly different spin on the project of investigating when we read, for they suggest the explanatory power of timing in an equation where sustained bouts of reading must compete with other activities that make up our lives.

In trying to answer these questions, I turned for a good while to book history alone. But the time reading takes is difficult to parse historically. Most studies of books and their reading map the traffic and dissemination of texts positively, connecting and correlating events of a book being printed, or bought, reprinted, or read. They show where books happened, not where they got stuck in the process of communication. Important studies in this vein grapple in different ways with questions of which books traveled, of how texts were shared and accessed, and of what influence they had as connective objects.[31] In concert, scholars of book and print history have built up an increasingly nuanced understanding of the way books spread in the later eighteenth century, and of the demographics and topographies of their reception within and between communities. But precisely because the authors of these studies investigate the publication, sale, or reprinting of books as historical events, they are less attentive to reading as something conjugated in time: to the sequences in which books (read fast, or slow, on repeat; as an alternate to not reading, or to working, or across a lifetime) show up alongside other objects and activities that take time. For good reason, histories of reading tend not to be histories of *not reading*—and thus to miss the very kinds of alternation and recombination and variations in time that I want to make visible as a property of book use in all its delights and its limits.

This bias toward reading as a positive event is even more pronounced—again, for good reasons—in digital humanities projects that aim to make the circulation of books visible by revealing networks of correspondence, or clusters of word combinations recurrent across texts far removed from each other, or collections of books distributed in space. The evidence provided in such projects can be compelling—but circularity also sets in. Digital representations of literary history, designed as they are to make connections visible, tend to do just this. Space matters more than time as the field in which visualizations operate most easily, and things like the pace at which a book is read, or the intervals between its readings, are much harder to diagram. Methodologically, an unread book can often register in databases about the spread of books in the same way as a read one. Digital research methods are now contributing at a rapid rate to our sense of eighteenth-century print networks as spaces of continuity and mediation, but as they show us how books move with illustrative bursts of color and bold lines through time and space, they neglect the links that weren't made—the breaks and spacings-out in time that would complete a picture of reading as something that spreads out in and defines time just as unevenly and pervasively as it does space.[32] Books, these approaches tell us, connect us. They connect us to each other, they connect books to other books, they connect distant times to recent ones, and they show objects connecting to subjects. Reading them brings us into line with each other. Even when we don't know that they connect us, the right algorithm proves that they do. But what of the *read* book, the book as it comes to life only when it is given a particular kind of attention? Doesn't the *read* book perforate as much as it connects—by being there sometimes and not others, or by vanishing and changing over time, even when the book itself sticks around?

In trying to imagine a more complete model of book use, we might learn from the field of anthropology that introducing time to the way objects are imagined in space changes fundamentally how a network is described. The discussion of the gift, taken up by Claude Lévi-Strauss, Marcel Mauss, Pierre Bourdieu, and Jacques Derrida, illustrates this by pointing out that an event's directionality and separability in time define giving as a practice. Referring to Mauss, Derrida writes: "The difference between a gift and every other operation of pure and simple exchange is that the gift gives time. *There where there is gift, there is time*. What it gives, the gift, is time, but this gift is also a demand of time. The thing must not be restituted immediately and right away. There must be time, it must last, there must be waiting without forgetting."[33] There is, in other words, no object that can be called a gift unless we look at where it stands temporally through the act of its giving. For Bourdieu, the material point is that acts of giving are always alloyed by occasion and opportunity.[34] For Derrida, this giving and taking of time links the gift to the poetics of narrative, opening up new ways to think about the timing and duration of a story. But both arguments suggest why those of us doing book history and material cultural studies might approach the circulation of books more theoretically. They illuminate the way a *read* book requires the cadence Derrida describes in relation to the gift as "a delimited time, neither an instant nor an infinite time, but a time determined by a term, in other words, a rhythm, a cadence."[35]

This discussion of the gift overlaps with another conversation in France, between Michel Serres and Bruno Latour. Serres, speaking to Latour in a conversation published in 1995, when Latour was most committed to actor-network theory, warns against using networks as models of the material world on grounds very similar to the ones on which Bourdieu warns against abstract models of exchange. "Networks," Serres argues, "leave an image that is almost too stable. But, if you immerse it in time, this network itself is going to fluctuate, become very unstable, and bificurate endlessly."[36] Serres appears here as a champion of non-chronological thinking. In opposition to actor-network theory, he describes his own historical method in terms of the crumpled handkerchief whose material points only meet in time, not when it is laid out flat in space. His *The Birth of Physics* opens with the image of laminar flow, in which atoms, falling though space, would participate at the same rate were it not for their potentially uneven relationships to time.[37] Serres's invitation to think of the network "immersed in time" is a challenge Latour takes up in *An Inquiry into Modes of Existence*. But it applies more immediately to the task of thinking of the material presence of books as one that is destabilized by thinking about the timing of their reading. Once immersed in time, a network of books would have to show the capacity of reading to pull certain seasons, or evenings, or distant lifetimes out of chronological time and to put them together like the points in Serres's crumpled handkerchief.

Taken more politically, Bourdieu's and Serres's comments on the way networks must be subjected to temporality can also serve here as a reminder that the practice of textual engagement always has loomed (and still looms) large for most as a question of having time to read, and of that time having become one of the things that modernity has not yet straightforwardly or equally given us. Models that visualize reading, even from a politically progressive point of view, as creating a stable point of connection between the reader and the book are often insensitive to a reality that regularly parts, not just the reader from her book, but also the reader and her book as a dedicated coupling (for instance, in college, on holiday, or on Sunday), from the reader and her book at other times of the life cycle and the week, where books may cross her path or populate her world to quite different effect (as paper, gifts, or guilty pleasures).

If we think of books simply as things fully activated as soon as they are on the move, we miss, as Serres points out to Latour, the very dynamic of the game: the way, as Serres puts it in this conversation, that the network becomes a soccer game in which only the ball, and not the players, move.[38] The challenge here is to think about reading as a practice that makes the book into an object that takes, gives, and occupies time unevenly. Visualized in this way, a book, like the ball that passes from one player to the next, will be held by some players differently at different points in their own game (or week, or year). As a thing with which they sit still and read for half a day, it is a very different thing from one that is opened quickly, or lifted and dropped or contemplated as something whose time is still to come. As a thing read long after it was written, or saved for later, it points in the network to time as a dimension of reading that cannot be easily mapped if we think only of books as objects. Differences in how one uses the ball in time introduce a site of distinction that may be greater than the connective feature of different people having held it or passed it on.

This is where systems theory comes into this project. Luhmann's emphasis on time is different from Serres's but is equally strong. His systems are constellations that make modern life fathomable by slicing it up into arenas such as love, law, art, and media, entities that constitute and maintain their character over time and by having different rates of operation. To work by analogy, these systems operate like sentences in a sea of words or bodies in a universe of atoms. They are engines of difference-making that create insides and outsides where no such distinction exists at the level of their parts. The components on the inside and the outside of systems are essentially the same: they could have been arranged differently. But once things are on the inside of a system they are defined by its terms—and these centrally involve timing. While Luhmann's law system, for example, involves the slow process of returning to precedent and performing deliberation, his love system involves the strenuous cultivation of the idea of constant and absolute communication (so that a fight, for instance, or a silence can occur within that system as communication). And while the media system involves the fast pace of constant feedback, the art system exists as a spacing-out and returning to objects singled out for perception. "Reading texts," Luhmann argues, "takes time—whether in narrative one reads the sequence that unfolds in the succession of sentences, or whether, as in poetry, one misses what matters if one thinks reading must begin at the beginning and end at the ending, and one will then have understood it all. When reading, and even more when looking at painting and sculptures, the observer is relatively free to choose the sequence of observations, so long as observational operations are arranged sequentially."[39]

Luhmann is suggesting that even selective reading of, say, non-sequential pages in a book becomes its own sequence in time. On this basis he can point to why a single text might appear in one system as informational or erotic and as art in another. A text that is not given time might still be read, but not with the duration he emphasizes here as belonging to the art system. Because there's no integral component of Luhmann's systems—no text, no infrastructure, no kind of person belonging inherently to one system rather than another—they model in surprisingly neutral terms the difference that timing makes.[40]

Consider, for example, a story that was reported in the papers a few years ago, of the world-famous violinist Joshua Bell, who appeared as an ordinary busker in the Washington, DC, Metro system.[41] Very few people, the *Guardian* reports, stopped to listen to him playing in this setting: commuters rushed by, intent on their journey, oblivious to the art hailing them to stop. In this reporter's terms, the event becomes evidence of us all having become Philistines, inured to beauty. But from the perspective of systems theory the outcome is obvious: of course art requires its own system, its own timing. There's no level of skill essential to a musician or a listener as effective as the system in slowing us down to listen, or to read, or to look when texts are being presented as art. No text in these terms is good enough to draw us in anytime and anywhere, and no reader is good enough to recognize a good text anywhere and at any time. What matters more is the quality of time it is given: the way a book and a reader, or a piece of music and its audience, interact under specific conditions where the system calls the tune.

I'm seizing here on just one aspect of systems theory. Others, specifically Luhmann's account of contingency, will come out in the chapters to come. But I want to address up front the most difficult, and the most helpful, aspect of using Luhmann in a literary historical study such as this one. The most obvious problem is that there are no people in systems theory. As my example of the busker shows, in systems theory, differences between systems are much stronger markers of meaning than any version of ideology or personhood or communication or content that might translate across spheres of modern life. If I work in a university at one point during the day, and become a patient in a hospital at another, nothing I carry over between these roles trumps the way two systems play out through me as differences. The difference between being a patient and a lecturer is greater than the sameness that comes with being me. Thus, while there *are* people in this book—Elizabeth Carter, the brilliant classicist; William Godwin, the principled radical; Thomas Turner, the Sussex grocer; Elizabeth and Richard Griffith, struggling writers and editors of their own correspondence, to name a few—my focus is on the reading systems (not the networks or the archives) that make them readers. And while there are also books in this study, including many significant enough to warrant close attention—Fielding's *Amelia*, Richardson's *Sir Charles Grandison*, and William Godwin's *Political Justice*—here their content matters mostly as a comment on the way a book passes through time as reading.

In other words, my use of Luhmann is in evidence in the emphasis I place on character and content as a matter of timing. While a conventional history of reading might look at the kinds of texts being read or the profiles of those doing the reading, Luhmann's arguments have allowed me to cut a path through my historical material that prioritizes time use, and the situating of the reader in time, over either of these factors. Luhmann once caused consternation by beginning a lecture with the claim that people do not communicate: "Only communication," he explained, "communicates."[42] The equivalent claim here would be that people do not read: only reading reads. We become literary readers, I maintain, by using books under certain temporal conditions: conditions that are sometimes partly, but rarely entirely, of our making. At other times, we are different kinds of readers, or no readers at all. We might be the owners or purveyors of books—of long and literary books, even—but we are not always their readers. Historians often find this distinction difficult to make, but I lean on it here as a way to write the history of a temporally discontinuous practice of reading, rather than of readers or texts. It is through this lens that reading becomes thinkable alongside eighteenth-century time-discipline as a practice that delimits and complicates, as well as establishes, the relationship of the eighteenth-century subject, and the eighteenth-century text in time.

To those who object to Luhmann's anti-humanism, I can only say I have used systems theory perversely in negotiating some of the most human aspects of my own life as a reader. At the time I was reading Luhmann, I had just given birth to my second child. I had to be apart from my family for fairly long bouts of time as my partner and I negotiated jobs in different countries. Thus, if ever there was a creative appropriation of Luhmann, it's mine: I read him in part as describing how those months felt to me, sitting at one moment in a library reading the driest of academic German scholarship, and at another pushing a stroller through windy streets in a wintry city. There was library time, baby time, and other times in which friends and students brought me joy. Books were all around me, and they figured largely in my fantasies about the future, but my reading was intense and erratic. The difference between my worlds of reading and not reading was so great that I was willing to concede that connection between them might be impossible. Giving up on identity as continuity did not feel hard. Moving between lives, I felt Luhmann, the least humanist of them all, inviting me to think about some of the most intimate strategies that people, women in particular, have used in partitioning their days, weeks, and lives, and making their reading fit those partitionings. Re-appropriating through book use the hours that labor and duty take does not, I've come to think, depend nearly so much on the maintaining of a steadfastly bookish identity as it does on the kinds of pasting, slicing, sequestering, poaching, and suturing that reading as a discontinuous and difficult-to-realize activity allows.

This brings us to the materiality of the codex book. To the extent that media are central to Luhmann's systems, they are ways of locating self-reflection within a system, rather than at the level of immediate human understanding. Love, he argues, develops by way of the printed romances and narratives that have reflected and enabled the development of the particular code of romance.[43] But my sense of how the printed book matters to eighteenth-century reading, particularly in facilitating the kind of reading associated with having and finding time, with brushing clock time against the grain, is more material than this. I am interested quite literally in the form of the book: its bound pages, its longevity, its shape as something that can be picked up and put down at discrete intervals.

All the chapters of this book, but particularly the last two, highlight these practices in their eighteenth-century contexts. In this domain, I have drawn on Bruno Latour's idea of "fiction" as a materially defined "mode," using his theory to supplement Luhmann's account of the temporal specificity of the art system. In respect to reading, Latour would point out that however much time a reader has to give, there is no reader without a text. The book reader can be seen in these terms as a historically specialized hybrid form of life, a body engaged in holding and turning pages, restricted by each book to reading a single text physically present in advance of the reader. A reader engaged in picking up and putting down a book, accessing with fingers its movable pages, is distinct from the reader of digital texts or phone messages, who can pick up and put down texts while holding a screen. In Latour's terms, finer distinctions could be made between the different practices of screen reading and, if we went further back, between readers whose bodies were engaged in unrolling scrolls or handling manuscript. Latour, like the media historians I'll turn to in a minute, would endorse the observation of all this particularity as an extension of Luhmann's idea that only communication communicates. Not only is it only reading that reads; it's only certain kinds of text format that facilitate certain kinds of reading.

Latour's monumental *Inquiry into Modes of Existence* converges with Luhmann's idea that modernity has been characterized by a plurality of different ways of doing things rather than by subjects whose practices and ideas are the work of a single consciousness. Latour's "modes" rely much more heavily than Luhmann's systems on material engagement: they are ways of using things, rather than using time. If Luhmann's law system requires as its process the slow pace of things characterized by Dickens in *Bleak House*, the mode Latour calls law requires judicial uniforms, papers, rooms in which certain kinds of events are to be staged. But Luhmann's art system and the mode Latour calls fiction converge in the sense that both describe certain ways of using a certain kind of thing (a book) in time as primary to the character of our interaction with it. The beings of fiction, Latour argues, arise when the raw materials of page, stage, or screen are treated differently from the way they would be treated in fact-seeking modes. Fiction is "a mode of existence like no other, defined by hesitation, vacillation, back-and-forth movements, the establishment of resonance between successive layers of raw material from which are drawn, provisionally, figurations that nevertheless cannot separate themselves from this material."[44]

Latour's celebration of this kind of activity arises from his observation that the handling of fiction's objects involves a path of deliberate return, a dedication of time to the book, and a repeated set of actions in relation to it, which follows from the awareness that "you have to keep holding it so that it will keep holding you" (247). It's not only that the world-famous violinist must have time dedicated to hearing him if his talent is to register; this time in Latour's terms would also have to be given like a gift, as a consciously dedicated and cultivated habit of listening to his music. In the wider sweep of Latour's argument, fiction becomes, like religion, a meta-mode. With it comes the high awareness that actions such as going to a concert, a church, or theater or reading a book matter *as a giving of time*. In his terms, fiction and religion are the modes in which the intertwining of objects and people is most fully acknowledged as needing duration.

Adding Latour and Luhmann to Thompson and de Certeau does not, then, in the end move us so far from that view of the literary reader as the one who resists the pace and logic of modern time-discipline. All four would agree that there's a slowing down, a repetition, a promise, associated with book reading that pulls back on the logic of accumulation and acceleration, and the measurement of time. But keeping Latour and Luhmann in mind turns attention away from the questions of who reads what and in what genre. The "when" of book use becomes as constitutive of any anti-capitalist character that we might want to claim for literature as the content or location of books: "One reading differs from another less by its text," states Borges, "than by the way in which it is read."[45] The most contentious version of this proposition would be that anything read in defiance of clock time, without being read for duty or for financial remuneration, and yet involving a conscious giving of time, acquires literary character. In the end, I will argue, it's as a way of being in time rather than investing or spending time that humanities education holds out its real promise. Literature as a use of texts in time, rather than a kind of text, or a kind of reader: to the extent that this is a study of literature, this is its starting place.

MEDIA HISTORY AS LITERARY METHOD

When did eighteenth-century readers read? Aren't there historical answers to this kind of question? There are, of course. Although reading is not the kind of thing that leaves an easy trace, one can find evidence of its having happened. My debt to histories of reading may not always be obvious, but this study is informed by scholars who describe in site- and medium-specific terms how reading has happened in the past. My interest in reading began with an admiration of Michael Warner's description of Mary Rowlandson's religious reading and with my graduate student self's discovery of John Brewer's description of Anna Larpent's reading in the eighteenth century. That discovery sent me to the Huntington Library to look at Larpent's journals, only to realize that Brewer had described them better than I ever could. My interest in reading was also fed by my fascination with Janice Radway's descriptions of women reading romances in the midst of family life.[46]

I have, of course, produced some historical descriptions of reading in this study that answer empirically the question of when people read. Diaries and letters will tell us, for instance, that the young Samuel Johnson read when he should have been at church; that for most of her long career as a mother, Hester Thrale struggled to find time to read at all; that William Godwin read in the very early mornings, in time carefully designated to books before breakfast; and that the young Thomas Holcroft read in snatches the books and posters he found on the walls of cottages or alehouses.[47] William Cobbett writes famously of a setting in the 1780s when "I had no moment of time that I could call my own; and I had to read and to write amidst the talking, laughing, singing, and whistling and brawling of at least half a score of the most thoughtless men."[48] Elizabeth Carter read by keeping herself awake through chewing on snuff and drinking green tea.[49] Edward Gibbon, barred during his time in the military from the life of study that he preferred, "snatched" his time with books, reading in his tent "amidst the tumult of Winchester camp."[50] Richard Griffith, a small landholder, read while supervising agricultural workers on his farm.[51] Library registers, like the stunningly complete one at Innerpeffray Library in rural Scotland, will tell us more generally that rural people read much more in winter than in summer, and tended to take out books toward the end of the week, even when the library was open on Monday and Tuesday.[52] Now Amazon and Google have begun to gather data that tell us exactly when we read: to anticipate, even, what we might read next. Empirically, one might say, the case is almost closed.

But I never set out to write a history book. For a historian, reading is what happens downstream of the text's production. Its reception can happen in surprising ways, in obedience or disobedience to the commands of a text about how or when it should be read. Reading can involve fans of Jane Austen getting her novels right or wrong, or collectors of books fetishizing certain editions, or Thomas Turner failing on every count to understand what *Tristram Shandy* is really about.[53] In my case studies, even though I am less interested in what people read than when, this obviously comes into play. But I am just as interested in the way texts anticipate reading as they represent their own continuation in time. An author who imagines the existence of her text has to imagine it being read or neglected, through intermittent perusal or its compulsive consumption. Reading is the temporal axis along which texts project their own existence.

This way of thinking about reading also makes a difference when it comes to my thinking about Latour. Were it not for my interest in what texts say about their own reception in time, this might truly be a study in the connection between readers and books. As I was writing this book, a friend suggested to me that I wasn't really writing about books. What about music, or art, or gardening? What about carpentry? Couldn't any of these ways of paying attention to objects be understood as ways of occupying time in all the iterations I was interested in? Carpentry as a hobby, for instance, also requires intermittent periods of deep attention. Over a lifetime, it might come into focus as a skill. And, to anticipate my last chapter, it might also help one foresee a future period of leisure when one would stop work and carve wood all day long. In this sense, I could as well be writing about woodcarving and the uses of time. To some extent, this is true. But the analogy works only if we accept that a woodcarving might have something to say about the way it was made and could be used in time. And it is to texts, not to wood, that I want to attribute this capacity.

To look for reading as something about which texts speak whenever they speak of time is to be drawn out of the domain of object ontology and of reading history and into that domain of theory in which post-structuralists once asked after the time of writing. When was this written? I am writing it now. For Derrida, that answer, written down, introduces a deferral, a form of difference that it became the job of the late-twentieth-century literary critic to unpack. Post-structuralism made us aware of writing as the addition of time to language, insisting upon it as a category that opens up words to an unknown future, unmooring them from their referents and from their origins in the mind of the writer. To write "it is nine o'clock" is to expose this statement to time as the thing that makes its truth empirically unavailable. This makes writing legible to literary critics in ways it is less obviously so to historians.[54]

The days when writing occupied us in the literary academy were also the days in which many of my generation were becoming typists and processors of words on our first personal computers. The focus on writing as something other than natural human expression made good haptic sense as our documents and digitally saved texts emerged for the first time as things to be written, stored, archived, copied, scanned, co-produced, distributed, and digitally published. It follows from this that we are now much more interested in how that volume of digital text and archive we've produced is to be read. Nothing has changed the nature of literary studies so much in the last decade as the new practices of digital reading. But that change alone is dramatically altering the way humanists think of their object of study, which is rendered new, even without addition, because of the different speeds and scales at which it is being read. How fast are you reading this? Are you a machine now? Literary historians of every period are now recognizing that reading, like writing, was always medium-dependent, and never self-evidently human.[55] In looking back to the eighteenth century, I am looking back to a moment when the reading of the printed codex book seemed most natural. Books came into view as everyday property, things that could actually be read and not just observed from a distance, and reading them became closely associated with the development of subjectivity. The serialization of narrative was not yet dominant and the circulation of manuscript was becoming more restricted. But it was also at this moment that reading began to take stretches of time that seem from today's perspective less and less self-evident.

Media theorists have expanded and made more concrete what once seemed like the post-structuralist project of mapping the relationship between the ways we read and write and the modern self. Friedrich Kittler offers a long historical view of the relationship between media and human evolution, situating writing within a larger media ecology where it stands out as belated to the event it describes. Writing, in Kittler's terms, can never participate in real time: its temporality is always symbolic (linear, alphabetic). This puts written communication at a disadvantage against the "technical" media that appear by the end of the nineteenth century as ways of capturing and participating in time itself, allowing us to rewind and replay events ostensibly without symbolic mediation. For Kittler, the eighteenth century is the last era in which writing could reasonably be read as direct human expression. Readers of *Clarissa* and *The Sorrows of Young Werther* might have felt the immediacy of voice in these texts, but henceforth, the reader of novels was destined to feel deprived of that immediate feedback loop that gramophone recordings and newspapers supported. The typewriter's arrangement of letters, words, and pages allows writing to continue, but only as constitutionally unplugged from the time of real experience.[56]

Under these conditions, most media historians would now agree, writing emerges as a historical medium and expires as the medium of current communication. Live transmission, writes Bernard Stiegler, brings with it "an exit from the properly historical epoch, insofar as the latter is defined by an essentially deferred time—that is, by a constitutive opposition posited in principle, between the narrative and that which is narrated."[57] The sinking of the Titanic or the landing on the moon become founding moments in a media historical process that foregrounds the role of communication but circumvents writing. The end of writing and reading as "live" occurs in Kittler's account later than the eighteenth century. But it is applicable in part to the perception of book reading in the period. For the reading of books was already opposed throughout the century to the "quicker" and more appealing consumption of pamphlets, periodicals, newspapers, and letters, suggesting the lag that makes them in media historical terms impossible to livestream. It is the same lag in which book readers with their hunger for time are arguably most at home, and the same lag that makes book reading easily seen as a recessive and privileged form of occupation.

James Lackington, writing in the later part of the eighteenth century, describes the difficulty of finding time for books. Lackington was founder of the London bookstore, The Temple of the Muses, which stocked and displayed at its high point over 100,000 remaindered and second-hand books. Images of the shop, described by his nephew as large enough that a carriage could ride through its aisles, have become popular as illustrations of late-eighteenth-century bibliomania, of a city drowning in books.[58] In 1791, Lackington published *Memoirs of the First Forty-Five Years of the Life of James Lackington*, an autobiography in which he gives his own upbeat account of this new climate. Most famously, and least creditably, he describes:

> The poorer sort of farmers, and even the poor country people in general, who before that period spent their winter evenings in relating stories of witches, ghosts, hobgoblins, &c. now shorten the winter nights by hearing their sons and daughters read tales, romances, &c. . . . If John goes to town with a load of hay, he is charged to be sure not to forget to bring home "Peregrine Pickle's Adventures;" and when Dolly is sent to market to sell her eggs, she is commissioned to purchase "The History of Pamela Andrews." In short, all ranks and degrees now READ.[59]

Promoting his own rise from Methodist journeyman shoemaker to autodidact and businessman of the new age with similar enthusiasm, Lackington attributes his own success to his powerful attraction to books, which he claims to have loved since childhood. "In reading," he exclaims, "I have experienced many thousands of happy hours, and which still engrosses the largest portion of my time" (453). Unpacking such boasts, and the version of self-help they help promote, underscores why post-structuralism and media history provide approaches to texts that are vital in supplementing those of the social historian.

At some level, Lackington's *Memoirs* tells a typical enough story of a young working man making the time to read through extraordinary feats of discipline. His tale resonated, particularly in nineteenth-century America, as a rags-to-riches story in which education plays an attractive part in bootstrapping the young book lover into his own story. But James Raven has argued with good cause that these *Memoirs* were always "difficult to take seriously" and "clear in their promotional aims."[60] This becomes evident if we look at the catalogs that Lackington oversaw from 1779 through 1793, thick books that circulated in order to advertise his proliferating stock in lists that grew from twelve thousand in 1779 to over twenty-three thousand in 1796.[61] Lackington valued the making of these catalogs. In the *Memoirs* he considers them his first publications and proudly describes writing them himself: "Indeed I continued the practice for years after my health was much impaired by too constant an application to that and reading; and when I was at last obliged to give up writing them, I for several catalogues stood by and dictated to others; even to the present time I take some little part in their compilation" (397). The catalogs were advertised and circulated, and there is strong evidence that they were read: a copy of the 1797 catalog survives as part of the King's Library housed at the British Library, while others held in special collections in states of unbound decay evidence how heavily they were used.

These well-thumbed booklets offer, of course, one version of the technical fix that Wellmon describes as a response to the increasing volume of print artifacts in the 1780s. Here, classified by genre and size, and listed with price and physical description, were the vast quantities of books on offer distilled into one catalog that could be searched at the flick of the fingers. Yet these catalogs also raise the question of time. Who has read all these books? Do so many books really come recommended by a single reader? Just compiling the catalogs, Lackington suggests, was exhausting enough. The figure of authority behind these annotated lists poses as a reader conversant with all of them. But this is surely a fiction—one that Lackington anticipates and supports when he addresses so carefully the theme of his own temporal economy. In this passage, for instance, Lackington describes his time as a journeyman: "In the winter I was obliged to attend my work, from six in the morning until ten at night. In the summer half year, I only worked as long as we could see without a candle, but not withstanding the close attention I was obliged to pay to my trade, yet for a long time I read ten chapters in the Bible every day" (100).

His superhuman efforts as a reader continue by his account into his later youth, when he shares books with a fellow workman in a system that allows them to maximize the hours a book was in use: "So anxious were we to read a great deal, that we allowed ourselves but about 3 hours sleep in 24, and for some months together we were never in bed at the same time" (165). And yet by the time we arrive at the period where Lackington is building up his own business, the account of when he actually reads has become more vague: "A bookseller," he argues, "who has any taste in literature, may in some measure be said to feed his mind as cooks and butcher's wives get fat by the smell of meat" (414). Perhaps this is true. But the only thing the bookseller is guaranteed to absorb by this method is exactly what the earliest surviving example of Lackington's catalog shows he has acquired: knowledge of the physical dimensions of his stock. With patience and adjectival excess, he catalogs stock "new in calf," "neatly bound," "calf-backed," "sewed," "gilt and rolled," and "new in board."

The idea that Lackington succeeds by virtue of his own learning, that he makes time for reading under impossible conditions, thereby becoming at later stages of his life a man of means, is a narrative of powerful mythical dimensions. And yet the more interesting story nestles literally between the lines of the hundreds of thousands of titles he lists in his catalogs. The catalogs from the 1780s organize and present their material, in some cases through generously long titles that disorder the page, but they give no recommendations to the novice reader about the material. Conversely, the catalogs from the 1790s, the ones published once Lackington's involvement in their physical composition was dwindling, are peppered with interesting recommendations on the stock described. His catalogs from 1792 include long titles and asides in the case of many items to a less experienced reader on the merits or demerits of each item.[62] The 1796–1797 catalog, for instance, recommends "Bannister's View of Arts and Sciences" by stating, "This useful, entertaining, and instructive Book, treats of the Architecture - Astronomy - Language - Mythology - Natural and Moral Philosophy of the Ancients, (same that is sold in boards at 3s 6d)."[63]

Lackington sold the Temple of the Muses in 1798, so there is no evidence that he was responsible for annotations in the catalogs after that point. But in the years between 1791 and 1797, it seems likely that his "little part in their compilation" consisted of inserting these comments. During these same years, he was revising his *Memoirs*, updating them for new editions that included references and quotations from contemporary authors, including Wollstonecraft and Wordsworth, authors he was evidently reading for the first time. *The Confessions of J. Lackington* (1804), the book he wrote after his retirement, testifies much more compellingly than the *Memoirs*, which he later regretted writing, to his daily life of reading: "I am grown more indifferent than ever to what others do, how they live, &c. or even as to what they think of me, or my way of life. I know that I am thought to be a strange sort of fellow, as I neither hunt, shoot, drink, or play at cards. I read until I am tired. I then walk or work in my garden, and in bad weather I cleave wood &c."[64] This description of retirement makes it seem likely that Lackington's last stretch of involvement with the shop involved reading the books he had been selling for years. The annotations that appear in the 1790s catalogs testify to someone familiarizing himself with the collection at an impressive but human rate.

It seems, in other words, that Lackington became in his fifties the kind of reader he'd said he was but actually wasn't all along. When do we read? We read when social, political, and economic circumstances allow us to open the books that have accumulated all around us. Those books may already have contributed to our view of the world without actually being read. Lackington's own policy of refusing to sell books on credit, thus issuing them only to those who already had the money to buy them, underscores the appeal of this sequence: first the money, then the outlay in education; first the books, then the time to read them. This may be the simplest and cruelest answer of where to find the history of reading's time. When are you reading this? Either for work or because you've organized in some other way the time free of work. But Lackington's story also shows why the temporal history of reading requires a literary and a theoretical rather than a historical approach. If its real contours are obscured by accounts like his, of boys claiming their portion of leisure through reading at unlikely hours stolen from backbreaking work, then taking texts at their word will not bring us closer to describing either the sequences or the hours it actually takes to consume books.

In terms of media theory, the withdrawal from calculable time that reading books requires becomes part of Lackington's autobiographical fiction. The books he describes reading in his youth are still there for him in old age: his reading of them can be inserted after the fact of their material existence in his world, and as props in his story. The speed at which their reading follows their publication is elastic and has no effect on their content. They can be read, as Lackington appreciates, after the fact. The temporal lag that comes, for Kittler, with translating things into and out of an alphabetic sign system becomes here the advantage of writing as a medium. This is the chronology that Lackington plays with as he reorders his life story to suggest that reading comes before the books he owns, and the hours of leisure stolen from the rest he will have come to earn. The very quality of the book's being independent of the event it reports on—of its being symbolic in Kittler's terms—may give us cause to hope that reading books remains viable and distinctive as a way of skewing and dilating time. Precisely because reading isn't a tool for making leisure time where there is none, and because books don't get into our brains magically (even when we work in a bookshop), they continue to require, but also to give, temporal non-equivalence as a precondition of their operation.

The chapters that follow concern the life of eighteenth-century readers, men and women whose reading I track by engaging their diaries and letters and by reading novels as things that anticipate their own uptake as something that will require the unevenness, the extension, and the anticipation of time. The theories I've named here as alerting us to the peculiar challenges of studying events in terms of duration rather than just as points in a spatial network will remain key. I don't have any general answers to give to the question of when we read, although some local ones will fall out along the way: unsurprisingly, it turns out, we (that we of literature workers that joins the then to the now of this study) read on Sundays and in summertime and when our paid work is done. But the more complex answers I offer concern time itself as something that is so interwoven with reading that the forms book use takes refer back to and allow us to catch hold of its more elusive states: to the lifetime, to free time, to parallel times, and to the future. These are shifty times. It's hard to say if we are ever in them, or if they ever really arrive. They are temporal zones that hover just beyond the horizon of homogenous-continuous time; never quite now, or then, or around the corner. As Muñoz has suggested of queerness as a form of temporality, they belong constitutionally outside straight time.[65] But it's relatively easy to say why we want them—perhaps now, more than ever—as sources of hope, of queerness, of resistance to the present, and of belief in the future. My job here is not to say how we will find them as twenty-first century users of media. It's only to show how book reading has helped us conjure time into being in the past—and to suggest that doing so has always been, as it must be now, a matter of political as well as personal struggle, and of creativity at the point of a text's reception as well as its invention.

The case studies that follow are organized around different forms of nonlinear time that emerge in conjunction with book reading. The next chapter features readers, most notably Catherine Talbot, distinguished in feeling the lack of time to read. These working men and women describe their daily time as broken up, overfilled, and frittered away. But they do not simply long for more leisure. What they desire more are protected hours, days divided more clearly between work (which might or might not involve reading and writing), and the kind of book reading that is emerging at this point as the complement to work. The reading they support seems to need this striation in time, but it is also instrumental in their own strategies for introducing books into their lives.

The second chapter is concerned with the role of repeat reading in the making of the lifetime. I look here at readers for whom the time of the happy life comes into focus through loops of return to a single text. Elizabeth Carter and William Grenville go back to critically reinterpret texts they know well, building up in the process a sense of life's happiness, not as mastery or progress, but as a way to recoup the time spent reading as visibly constructive without being obviously progressive. Books feature in their materiality as things that are stable even as readers are not, becoming touchstones in the process of reevaluation that this cluster of politically awake readers valued. While my focus here is not on novel reading, I end by looking at *Adeline Mowbray* as a fiction that shows Amelia Opie's support for this idea of the lifetime as the proper duration in which to read a book.

The third chapter directly concerns novel reading and is also the one most directly concerned with the shape of the codex book. Here, pages that are bound show up as things that can be read over time as out of order. In fact, their being bound is what opens them up to nonlinear reading. This is a model, I suggest, for the way Richard and Elizabeth Griffith, busy, self-educated professionals, read without feeling that they lack time for books. It is also a model for the way they represent their own marriage as something that unfolds chronologically (like a novel), while opening itself up to different views of how that time might have been spent. Reading's tense is connected here with contingency: using books as the Griffiths do, as things to be opened randomly and in no particular order, helps brings to life the multiplicity of what could have been, as plot and page help open each other up.

One of the main problems with both Latour's and Luhmann's social theories is that they make it hard to imagine directing our own future. Any utopian project must flounder in the terms of either modes or systems, neither of which can account properly for human hope or the creation of political horizons. The book's last chapter, then, turns away from these theories to suggest why books in their materiality have allowed us to invest in a future—a future in which the real time for reading is still to come and toward which the materiality of paper and print has always been able to point. My case studies here involve Elizabeth Inchbald and William Godwin, readers and writers of the 1790s who are invested in different ways in the belatedness of reading to the scene of revolutionary action. But I end here because this is one of the places where the nexus of reading and time-making I have explored comes into focus as most current. If the reading and stockpiling of books have involved our keeping an eye on a future in which there will be more time to read, where does their absence leave us? If eighteenth-century book readers imagined a time of reading still to come, are we that future—or did that project fail?

CHAPTER 1

Time Divided

> I find my thoughts, increasingly, not on the supernatural or spiritual, but on what is meant by living a good and worthwhile life—achieving a sense of peace within oneself. I find my thoughts drifting to the Sabbath, the day of rest, the seventh day of the week, and perhaps the seventh day of one's life as well, when one can feel that one's work is done, and one may, in good conscience, rest.
>
> —Oliver Sacks, "Sabbath," *New York Times*, 2015

NO DIFFERENCE

According to Jonathan Crary, our main problem today is not a lack of leisure time. Rather, it is the fact that we've entered the twenty-first century as beings radically unable to draw a line between leisure and work. We are always switched on. We are losing the ability to differentiate, Crary argues, even between waking and sleeping. What he dubs "the logic of 24/7" "disavows its relation to the rhythmic and periodic textures of human life. It connotes an arbitrary, uninflected scheme of a week, extracted from any unfolding of variegated or cumulative experience."[1] There's no hour when we can't shop, or work, or read the very latest news, or cruise the internet for sex. This condition of nonstop awakeness is facilitated in Crary's account by digital media: "Today, the permanently operating domains of communication and of the production and circulation of information penetrate everywhere. A temporal alignment of the individual with the functioning of markets, two centuries in developing, has made irrelevant distinctions between work and non-work time, between public and private, between everyday life and organized institutional milieus."[2]

There are, one could argue, other versions of synchronicity, habits associated with new media, some of which we are not yet even fully aware of, some of which are just coming into being as people learn to switch off in ways their parents can't fathom.[3] But Crary's claim, that current reading practices overlay the diurnal and seasonal rhythms print reading once helped us keep, is difficult to refute. More mysterious, however, is why this new culture of ubiquitous online reading should threaten so directly the reading of books, a form of textual engagement that was never as closely linked as serial reading, postal communication, news consumption, or television viewing to the striation of time. Or was it?

I might just concede that online reading has colonized hours I once had for novel reading. It's not necessarily that I have less time now than I did twenty years ago. But I don't have the evenings that came when communication could be turned off. Instead, I have the endlessly updated newspaper websites and emails, many of which invade my nights as missives from other time zones with the appeal of being live, or being just for me. They can be read right up until I fall asleep, in the very hour that I once reserved most stubbornly for books, and they are there in the morning, in the time I once used to dive straight into writing. Nor do my children have the holidays of my childhood, in which I often drew a book into bed in the morning and wandered in the afternoon through a stack of public library shelves, collecting books randomly and reading them until late. Instead, it seems fair to say, a constant tumult of news, correspondence, and Facebook posts makes even holiday time feel broken up.

To admit to the vulnerability of my own time for book reading, however, is not necessarily to locate myself in the twenty-first century quite as squarely as Crary imagines. Since at least the eighteenth century people have worried about a state of constant distraction that works against the reading of books, a failure to find stretches of time in which to settle down to read as one hopes. Many people never had those summers or evenings I've just described. Others who might have had them saw themselves lacking the self-discipline or structure needed for scholarship. Even in the 1700s, most were less sanguine than Lackington about reading as something that book lovers would stubbornly find time and opportunity to do. Eighteenth-century readers write of it being hard to sequester time from work and of the feeling that more immediate or compelling kinds of reading interfere with the real reading they've resolved upon. Women in particular complain of the buzz of social obligation and communication from which they find it impossible to protect themselves.

In the 1750s and 60s, when Catherine Talbot writes to her friend, Elizabeth Carter, she complains routinely of reading and writing being squeezed out of her days by constant duty and small-scale distraction:

> I am ashamed of myself, dear Miss Carter, but my time goes at a suprizing rate, and nothing seems to be done. A long ride sweeps all the morning, and then I

> prowl about the garden. After this I determine, like Phyrrhus, to sit down and enjoy myself, write, read, and be quiet. But then Anne Such-a-One has brought a pound of spinning, there is a new brood of chickens just hatched, Phillis is ill, powders must be weighed out, or Shaw consulted; Flora wants to talk to me; Polly and Dicky are good children and I must gather a posie to reward them.[4]

During a correspondence that proves how much she read and wrote despite these duties, Talbot worries constantly about her lack of time and failure to read books deeply. I begin with her complaints in this chapter because they suggest where Crary's diagnosis of the present misses something important about the relationship between time and book use in the past: for readers like Talbot there has never been an easy way to draw a line between work and leisure. Book reading figured as a casualty of that blurring, as well as being a reason for people wanting to shore up the work-leisure distinction.

Talbot's response to her own predicament is to try to protect Sundays as a day of study, a strategy other eighteenth-century readers in this chapter also use in making a temporal reserve for books. My concern here is not so much with the Sabbatarian cause they support, or with the debates around what kind of texts should be read on Sunday (although I will touch on these topics), as with their desire for a mode of reading that inherits from religion the language of temporal differentiation. The readers in this chapter make bouts of uninterrupted and sustained reading seem vitally dependent upon the dedication or giving, to use Derrida's term, of special kinds of time to books. Like the music of the world-class violinist, inaudible as art when one catches it by the snippet on the way to work in the subway, they approach what we might now recognize as literary reading as something that must have its own temporal coordinates. Sundays are an obvious eighteenth-century example of when books might be held in that attentive way, but winters and periods of youth might serve as other case studies.

Generally, then, this chapter follows up on Luhmann's description of the art system as one that requires duration, and Latour's of fiction as a mode characterized by a certain dedication of time rather than by its specialist materials. Fictional and factual narratives, Latour stresses in *An Inquiry into Modes of Existence*, "are made of the same material, the same figures. . . . And yet, starting from the same basic raw materials, the two modes differ through the treatment to which we subject them" (251). While "fiction" carries us away, fact-seeking modes see us "disciplined by chains of reference" to the "real" world for verification. The distinction between these modes is in the passage and the disconnection between them, which fiction supports by being a dedicated time—but one we cannot stay in. We must leave it and come back to it in a movement that is habitual but not constant (251–53). For Luhmann, in *Art as a Social System*, this same observation pertains to the art system, within which he describes the artwork appearing as part of "a temporally abstracted structure. It is a program for repeated usage which—like today's complex computer programs—blocks access to what is going on during the execution of the operation" (43). But one cannot stay in the art system any more that one can stay in the mode of fiction: in Luhmann's terms, art emerges through improbable combinatorial structures "wrested from everyday life" and designed for intermittent use (128).

These arguments resonate with how the readers in this chapter wish, not for lives of leisure that would allow them to read constantly, but for more compartmentalized lives—ones that would allow them temporally designated zones of intense engagement with books as an alternate to work, and to other kinds of more extensive or instrumental reading. The Sussex merchant Thomas Turner, for instance, uses his diary and account books and does ample reading throughout the week but wants his Sunday reading to be different in quality and quantity.[5] Vicesimus Knox imagines a reader who will retire with books on winter evenings once he has disengaged from the news and the live and theatrical entertainment of the summer, absorbing them then by shutting out the wider world and its print ephemera temporarily.[6] In *Memoirs of My Life*, Edward Gibbon pinpoints an intense period of his own study in Lausanne, between 1756 and 1758, in which he settles into habits of reading that elude him later, and which he struggles to recreate (34).

Both Latour and Luhmann would stress that these are cases in which there is no directly causal relationship between a certain kind of reading and a certain kind of occasion. It's not because Turner reads special books on Sunday, or Gibbon reads better texts in Switzerland, or because of when or how these texts were published, that books come into special focus for them. That explanation can be applied to good reading, of course. We can say certain kinds of texts attract certain kinds of attention. But the evidence in this chapter suggests that readers like Talbot, as well as Samuel Johnson, William Temple, and Thomas Turner, value a relation between reading and the division of time that is not primarily about content. The dedicating of a day or a term is important to a certain deep reading, which feeds back in turn into how one feels about a text. This may seem tautological, but it's part of the logic that pertains to the book as something that flares to life under special temporal conditions, just as it does under spatial or institutional ones.

At its most concrete this chapter is about why Sundays have been relevant to our understanding of reading as an activity defined by the dedication of time. At its more abstract, it is about why reading of the kind that we might now recognize as belonging to these modes or systems has depended on and reproduced temporal striation. I lean here on arguments that have presented modern time as more heterogenous than Stuart Sherman or Benedict Anderson suggest. Peter Osborne, for instance, contends that Anderson radically overstates the continuity of modern time-consciousness by falling for a narrative entirely internal to it: "When Anderson argues that the temporality of modernity knows no internal principle of variation, he is only partly right. He is right to the extent that the concept of modernity, in its basic theoretical form, itself furnished no such principle. He is wrong, however, insofar as it must find one elsewhere, if there is to be any way of identifying the historically, as opposed to the merely chronologically, new."[7] Recently, there been much more emphasis on the plural forms of temporality that modernity produces.[8] David Henkin argues for the rhythm of the week having structured nineteenth-century time-consciousness, particularly after the decline of religious and festive ways of marking days and seasons.[9] And Lloyd Pratt explores for the same period what it means to occupy multiple temporalities at once. In Pratt's terms, official versions of daily and historical time did little to coordinate their relation, even at the level of a small group, or a single text, for whom the future and the present could exist simultaneously in different forms.[10]

For my project, the more material question is if, and how, book reading has participated in this striation of time—even if texts themselves can be shown to contribute to its homogenization. Did the readers of books (as opposed to the readers of newspapers, websites, and text messages) benefit in particular ways from their days, weeks, or seasons being "wrest from the everyday?" Was this kind of differentiation a precondition, as Talbot's case suggests, for a certain kind of education? Did this mean that the life of constant leisure and the life of constant work were equally hostile to reading? From these historical questions comes the larger one of whether the reading of books as "fiction" or "art" contributed to the effect of temporal differentiation more than other kinds of media consumption. Whatever common ground I share with Talbot as a woman who never gets to sit quite long enough to read a book deeply, perhaps theater, newspapers, and books all helped to divide her time up, as Crary suggests, while my online reading just runs it all together? I end this chapter by looking at a novel Talbot knew and admired, *The History of Sir Charles Grandison*. Reading against the grain of arguments for the eighteenth-century novel drawing its reader into modern time, I introduce *Grandison* as a book that supports its uptake in the different kinds of time Talbot wants in her life: both that of the regular, daily time that Richardson's plots seem to register, and that of the isolated and sequestered times upon which Talbot feels her own best reading relies.

TALBOT'S LACK OF TIME

Literary history can seem full of women frustrated with their lack of time for reading. Florence Nightingale rails in *Cassandra* (1852) against the way women are constantly interrupted and never protected in their study, complaining that "there is no time appointed for this purpose and the difficulty is that, in our social life, we must always be doubtful whether we ought not to be with somebody else or be doing something else."[11] Virginia Woolf makes this frustration into the beautiful manifesto, *A Room of One's Own* (1929). Few, however, seem quite as angry about their lack of time as Catherine Talbot. She rages in her unpublished journals about not having enough time, she muses on her lack of time in her published pieces of writing, and she makes time a constant theme of her letters to Elizabeth Carter. As friends, Talbot and Carter had much in common. Neither married, both belonged loosely to what we now think of as the Bluestocking Circle, both knew Samuel Richardson and Samuel Johnson, and both were nourished to different degrees by their Christian faith. But Talbot's situation was particular because she grew up under the protection of her father's friend, the Bishop Secker, and was obliged to him for including her in his busy, affluent, and often very public household. The intensity of this situation comes out at one point when Talbot erupts in fury at Carter's failure to understand that her business is of a special degree: "You suppose that when I complained of wanting leisure I had several hours. You forget that you rise three hours earlier than I am allowed to do; that we visit eighteen families at from three to fourteen miles distant, and twenty I believe in Oxford, and are besides eternal riders, walkers, and airers. That I have many correspondents, and cannot for my life write short letters. And with all that crowded together, at first I had scarce <u>one</u> hour."[12]

Comparing her own days to Carter's more provincial and less privileged ones, Talbot describes a round of social and secretarial duties requiring her attention as adjunct to Secker's ecclesiastical role. As often as Carter urges her to see the advantages of social life in London, Talbot complains of the disadvantages of that life to scholarship. Her letters appear as written under impossible conditions: "I have absolutely no time," she begins one letter to Carter, "well, that is no matter, for positively you shall write to Miss Carter, before you are half an hour older—half an hour, why in that half hour I have half an hundred things to do" (1:174).

At various points Talbot rails as fully as any eighteenth-century Christian woman could against a system that precludes her from having what Nigel Thrift calls her "own time."[13] "How much rather would I stay at home this evening and study," she exclaims to Carter, "than to go out shivering in the cold and to pay half a score of unedifying visits" (2:199). Although Talbot was recognized early on as a literary teenage protégée and valued by Secker as an interlocutor throughout her life, she found it harder as she grew older to justify her education. And while she played key roles in literary coteries that Betty Schellenberg identifies as two of the most important of her day—the one that formed in the late 1750s around Elizabeth Montagu, and the one around Samuel Richardson—Talbot was a conservative producer of manuscript and wary of writing for even limited publics.[14] She complains to Carter that what skills she has cannot endure the life she's obliged to lead: "By some means or other my golden hours are all engrossed and I cannot help feeling a perpetual dissatisfaction—feeling that my little genius was not made to take in so large a round, even of proper and laudable engagements" (1:329). In 1758, Secker became Archbishop of Canterbury and his household moved to Lambeth Palace in Westminster, whence Talbot writes with increased frustration about the fact that her days have become even more strictly scheduled. Her evenings and weekends now offer more limited freedom: "I am glad enough in the evening of two or three solitary hours to read and write. Indeed, I seldom have as much, admitted into the study between eight and nine" (2:309).

Talbot's sense of failure to command her own time, and her tones of anger and frustration, support Rhoda Zuk's description of her "relegation to a round of undervalued employments." Talbot's "unhappy resignation to her role of modestly but variously occupied lady," Zuk argues, "paralyzed her confidence and even generated despair."[15] Sylvia Myers and Emma Major both see her in similar terms, as tragically interred in the work that comes with her role as an unmarried woman within Secker's family.[16] They remind us of the independence she lacks structurally and of the difficulties any woman faced, even in this relatively fertile period for women writers. While Schellenberg, who homes in on Talbot in some detail in *Literary Coteries and the Making of Modern Print Culture*, portrays her as a pivotal and highly intelligent player in coterie life, a mediator of others' relationships, she also singles her out as an ambivalent participant in these intellectual circles. Only Norma Clarke puts a slightly different spin on Talbot's intellectual life by emphasizing that many of Talbot's modes of complaint are quite conventionally Christian, and reading her as a productive sermonizer on themes such as time use.[17]

Talbot's frustration at not settling down to read is often directed at herself and displayed for an audience who shares the problem of self-discipline. Talbot's Christianity comes through as she berates herself for her failure to use hours better. She reports in her journal: "I feel the shortness of time most uneasily, certainly for no other reason than because I do not know how to make the best use of what I have, but cut it into snippets in the strangest way you can imagine."[18] Her journal describes her "strange careless way of using time" and "the witchcraft that makes all my time glide away without suffering me to do anything in it to any purpose" [5]. Her letters often express her guilt at having "as usual broke my time with a thousand little errands and employments that I did not make the improvement of it I might and ought to have done" (1:222). At her most pious, Talbot considers her failure to redeem the time she has on earth: "Has," she asks rhetorically, "my Time been improved or lost, or worse than lost, misspent? If the last, let me use double Diligence to redeem it."[19] And even when she has leisure, she resents her body's need for rest as much as she loathes social obligation at other times. "But alas how useless am I and how little improvement do I make of all this leisure," she writes in her journal of 1751: "How perpetually interrupted is the progress of the embodied mind too of so slight a Body, so liable to weariness, that makes such a large demands of time for refreshment and amusement" [22]. In a later letter she describes herself to an unknown correspondent in desultory terms, recognizing both her opportunities and her propensity to waste them: "Among sober people that have leisure to think and to employ themselves to purpose if they would I am certainly one of the very very lowest and most insignificant."[20]

It's worth noting that Talbot's days of work are not those of real laborer poets like Mary Leapor or Stephen Duck, whose careers she knew well. She is not a time-strapped reader like Thomas Turner, whom I discuss later in this chapter. The work that distracts Talbot is generally deskwork or socializing, much of which Carter registers with envy as providing her a life of constant interest. As secretary, scribe, and conversationalist in a household where reading aloud was routine, Talbot is actually in contact with texts during many of the hours she longs to spend alone with books of her own choice. While she wants to do more substantial bouts of reading, the "thousand minuntie [sic]" that take her time are often texts: "Bills of fare, messages, letters of mere business, are Sybil's leaves dispersed by the breeze of the every day" (2:19–20). On May 11, 1751, her journal describes a day spent writing "useful" letters, taking a long walk during which she reads Milton, and reading aloud after meals [5]. This is hardly a life in which she is deprived of books.

It might be useful here to compare Talbot to Edward Gibbon, someone whose gender, class position, and standing as a historian means that we think of him as having very different advantages than Talbot. On the face of it, he certainly had opportunities Talbot lacked: he was sent to university and inherited the profits of a large estate. But he loathed university life and struggled, both in his life in London and in his time in the militia, to find time to study: "My literary leisure was much less complete and independent than it might appear to the eye of a stranger."[21] As a soldier, Gibbon undergoes an eight-month period during which, "amid the perpetual hurry of an inn, a barrack, or a guard-room, all literary ideas were banished from my mind" (72). Upon his return to civilian life, he reports on his delight at being restored to the "pleasures of reading and thinking" after this period is over (72). In the bustle of London he was often without books, and while he had them at his father's estate in Hampshire, he complains of his days there, "I was not master of my time" (94). In his youth, the pleasures of his Hampshire library were mitigated by "the restraint imposed on the freedom of my time":

> By the habit of my early rising I always secured a sacred portion of the day, and many scattered moments were stolen and employed by my serious industry. But the family hours of breakfast, of dinner, of tea, and of supper were regular and long; after breakfast Mrs Gibbon [his stepmother] expected my company in her dressing-room; the newspapers; and in the midst of an interesting work I was often called down to receive the visit of idle neighbors. Their dinners and visit required, in due season, a similar return, and I dreaded the period of the full moon, which was usually reserved for our more distant excursions (62).

This comparison doesn't detract from Talbot's complaints, but it does help situate them in a context where even the most privileged readers felt the difficulty of spending long stretches of time reading alone. It also confirms Clarke's more general point, that when Talbot writes didactically of the desirability of redeeming time better, she touches on a topic that resonated with many whose conditions were quite different from her own.

Talbot's desire for more time with books did not amount to a wish on her part for the life of full aristocratic leisure. When she writes for a wider circle, her posture, at least, is as someone who values work, and sees it positively as a Christian obligation. "All persons," she writes in her *Reflections on the Seven Days of the Week* (1770), "should consider to whom they are accountable for their Time, their Labour, the Superfluity of their Fortune" (35). One of her *Dialogues* takes direct aim in this spirit at the woman of leisure, asking her what she has done with her summer:

> Rode, and laughed, and fretted.
>
> What did you intend to do?
>
> To learn Geography, Mathematics, Decimals fractions and good Humour: to work a screen, draw copies of two or three fine Prints, and read Abundance of History; to improve my memory, and restrain my Fancy; to lay out my Time

> to the best Advantage: to be happy myself, and make every body else so. To read Voltaire's Newton, Whiston's Euclid, and Tillotson's Sermons.
>
> Have you read nothing?
>
> Mrs Rowe's Works, the Tale of a Tub; a Book of Dr Watt's; L'Historie du Ciel; Milton, and an Abundance of plays and idle Books.[22]

Reading is derided here as a piecemeal activity, a form of distraction and frivolity, and the irresolute woman becomes the culprit in not having studied better. Her social condition as a woman may be at fault—Talbot undoubtedly sees women's lot as disadvantageous—but her own disposition and failure to structure her days seem almost equally to blame for her not becoming the reader she wants.

This has the interesting effect of Talbot suggesting explicitly at some points that the reader whose life lends itself best to proper study might be the manual laborer. Talbot often aligns herself with men of lower status on these grounds, envying them their slim but well-defined portion of a working week in which to enjoy books. *Reflections* channels the voice of such a worker, sympathetically presenting the man who has "scarce a Moment free from the necessary Engagement of business and bodily Labour. While I am working hard for Bread for myself and my Family, or attending diligently the commands of a strict Master, to whom I am justly accountable for every Hour I have, how can I find frequent opportunities for studying the word of God, or much Time to spend in Devout meditation?" (17). While the worker here seems unlikely to access books, Talbot implies that he may yet be a reader with real advantages over a woman like herself.

To this hypothetical worker Talbot replies that he may spend limited time with the right sort of book: short breaks here or there will "gladly" be spent studying the Bible, and Sundays are his own. Talbot takes up this theme in another essay, describing herself as a privileged woman on holiday, lying in the grass, observing a man employed in fixing a roof. Watching him, she is overcome with a desire to change places: "I would have resigned all these Delights with Joy, to sit whistling at the Top of a high Ladder, suffering both Heat and Hunger."[23] The man's advantage over her is that he has a real occupation with which he earns virtue. "Man is born to labour," Talbot exclaims in this essay, "it is the condition of his Being; and the greatest cannot exempt themselves from it, without a crime" (103). But having a job also liberates him to enjoy books and contemplation when his work is done. As Talbot's ideal scenario, this arrangement underscores what she finds so hard about her own life, which is that it leaves no time she can truly call her own. The presence of texts to be read and written throughout her day works against her desire to bring them intermittently into focus.

BREAKING THE WEEKLY ROUND

Sociologists have commented at length on the importance of temporal differentiation in modern society. Sociologist Émile Durkheim and anthropologist Edmund Leach have written, for instance, on the universal human need to establish alternates among existential domains. Durkheim approaches the division of time as a functional analogue of the division of space.[24] More recently, Eviatar Zerubavel has written about weekly time, suggesting that we depend, if not on the sacred nature of Sunday, then on the difference it introduces into the week: "Whether the discontinuity that the pulsating week establishes is one between the sacred and the profane, consumption and production, active socializing and isolation, freedom and obligation, the domestic and the public, or spontaneity and routine is rather insignificant. From an experiential standpoint, the most distinctive feature of this cycle is the fact that it helps to introduce discontinuity into our life and thus promote its multidimensionality."[25]

Looking back historically, David Henkin stresses the week having been basic to the experience of time in nineteenth-century America. He emphasizes the shifts during this period in "how the week was used to divide labor, schedule meetings, remember appointments, recall events, cultivate habits, arrange liturgy, maintain standards of hygiene, entertain strangers, communicate across distance, and mark the passage of time."[26] It is not so much that the week itself changed as a structure, as that its presence became more important. Certainly, in eighteenth-century England, the debates around how and if Sundays were to be protected from the encroaching hours of paid work were already energized by the consciousness of what Henkin calls the "hebdomadal cycle." Take, for instance, this 1753 defense of the worker's right to Sunday:

> Six days has man in duteous Toil employ'd:
> His Sum assign'd. And now the Eve appears,
> Prelude to sweetest Hours of holier Rest;
> Kind Respite, in the round of weekly Time,
> For travell'd Dust to call its Labourer home
> The partner Mind; to steal her from the Throng
> Of loud Intruders, charg'd with wordly schemes;
> And strike a partial Truce with mortal care.[27]

Arguments like this were being made in the face of the real danger that there might soon be only one kind of continuous labor interspersed with stolen moments of leisure. We have seen this possibility expressed in *Low-Life*, a text that shows a raucous population engaged in a range of activities on a Sunday, contriving, as the preface puts it, to "murder not only common Time, but that Portion of it, which is more immediately consecrated to the Glory of their great creator" (A2). The calendar that was in place for thirteen years in France after the revolution included no Sundays, confirming the possibility of producing what Sanja Perovic describes as "a homogenous rhythm ensuring that the only spectacle that would take place was that of living bodies coordinating their daily activities with one another."[28] This underscored a fear that had been mounting for some time in Britain, that there was nothing inherently sacred about Sundays. Unlike summers and full moons and evenings, Sundays were a human construction, one for which a case had to be made if they were not to become yet another workday in the new logic of capitalism.[29]

The arguments in eighteenth-century England about the importance of Sundays came from groups with distinct agendas. Sabbatarians invested in keeping the day sacred had a case distinct from those angry about workers being deprived of rest. The latter were invested in protecting leisure time, partly because this time could be used to organize workers as a lobby. But historical changes in patterns of work and worship affected both groups. Church of England services, which had long kept people occupied on Sundays, were becoming less popular, and attending twice on a Sunday had become much less common.[30] Declining rates of church attendance fueled fears of people drinking and committing other sins on that day, but they also signaled the danger that people might be tempted to see Sunday as just another day of work. The general winnowing down of leisure time in the period also fueled Sabbatarians' fears that what free time people had was less likely to be spent at church. In the mid-century, holidays still included All Saints' Day, All Souls' Day, Lord Mayor's Day, Prince of Wales's Birthday, Queen's Ascension Day, and the Powder Plot Day, in addition to which most people at this point still kept Monday as a holiday. Until late in the 1780s, Thompson reports in "Time, Work-Discipline, and Industrial Capitalism," "there are few trades which are not described as honoring Saint Monday: shoemakers, tailors, colliers, printing workers, potters, weavers, hosiery workers, cutlers, all Cockneys" (73). But by the end of the 1700s, most of these non-religious holidays had been eliminated from the worker's life. The majority of employees, for instance, in the new Wedgewood china factory had no Saint Monday and far fewer holidays.[31] As long as Mondays (and often Tuesdays) were unofficial holidays, and the calendar generously dotted with pagan holidays, Sundays could be preserved for worship alone. But in a weekly economy more tightly defined, it became a day on which a range of diverse activities competed palpably for time. Sunday in the newly conceived eighteenth-century calendar had to do double duty as the most significant stretch of free time granted to working people *and* as the day on which any kind of spiritual life was to be led. By the nineteenth century, Alexis McCrossen reports, "Domestic, didactic and commercial meanings for Sunday joined rather than replaced religious meanings."[32]

Sabbatarians and potential trade unionists were also both interested in making sure there was room for reading and writing during the working week—and in assigning a timeslot to those activities. Reading was something one might conceivably do instead of going to church—an activity, like walking, that might be used on Sundays to prevent drinking and debauchery. Sundays were also occasions for which special kinds of texts might be supplied. Evangelicals and old-school Anglicans alike began to endorse this latter possibility, and the booming publishing industry began to exploit it. Pious texts implicitly designed to be read on Sunday included works like *A Course of Lectures for Sunday Evenings, Containing Religious Advice to Young Persons* (1783?) and *The Family Preacher: Consisting of Practical Discourses for Every Sunday Throughout the Year* (1776), which addresses itself as a series of thirty-minute readings to those keen "not to squander away those precious moments in idle, if not sinful recreations, as is the customary practice of the present gay and licentious age."[33] Diaries were issued and kept whose format allowed for longer entries on Sunday, many texts designed for use in Sunday schools appeared, and appeals to Parliament to control drinking and traveling on Sundays were common.[34]

The only piece of writing Talbot published in her own lifetime, written in 1750 on the theme of Sunday, appeared alongside these texts in defense of keeping Sundays distinct. The short essay appeared as a letter to Johnson's *Rambler* and is written in the voice of a Sunday plaintively protesting its modern mistreatment. Talbot's Anglican plea on Sunday's behalf is a relatively modest one: she wants neither pagan festivities nor Puritan solemnities. Sunday as she personifies it wishes it "would every where be welcomed at a tolerably early hour with decent good-humour and gratitude."[35] It is a day for walking, clean clothes, and easy conversation. Above all, it is a day for reading: the piece ends by proposing that the *Rambler* publish a further essay on just that topic, advising its own readers on good Sabbath reading. The suggestion here is that the *Rambler* itself might fill that slot. Talbot writes at just the register she recommends for her Sunday reader, modeling in her "letter" the kind of text she expects to be picked up in that timeslot.

We know from Talbot's letters that what she describes in this piece is the kind of Sunday she personally relished. Talbot's investment in Sundays being different from other days goes back to her quest to claim her own free time, and to her own preference for a studious life. "London has its quiet hours," she writes to Carter, "for people who keep out of the impertinent racket of it" (2:64). In this same letter she describes going to church early, spending some hours in conversation, and enjoying Sunday evening as the time allotted by herself and her mother to staying home. She contrasts their position favorably with that of the "fine ladies" who are on their way to "drums and plays" as part of the Sunday crowd represented in *Low-Life*. This is the time of the week in which she allows herself to fall deliberately out of step with public life, conscious of Sunday as a day that offers her some protection from active duty. Her strategy doesn't always work, though. "One is always in a hurry," she writes to Carter: "Even of a Sunday I have folks to speak to, children to school, and many such matters to dispatch" (2:105–6). And Talbot regularly complains in her unpublished journal, either of weariness, or that "Sunday mornings are so much too short" [38]. But at Secker's country house Talbot often finds herself content. On Sunday, November 3, 1751, she writes in her journal: "I like Sundays here extremely. One has so much Leisure" [29].

Talbot's preference for Sunday as a day of retreat supports some of the more orthodox arguments for its being a day of sobriety and meditation. But Talbot is more invested than most in the fact that Sunday is a day on which to turn to books. Arguably, she's also far less invested in which texts are read than in the mode in which books will be handled once people are left alone to read as they wish. Her general assumption is that pious reading requires pious texts. But Talbot lets the emphasis rest on Sunday reading as a mode of activity rather than a genre of writing. In this she agrees with Dorothy Kilner who, writing as a mother in the 1780s, recommends reading and walking as the best Sunday activities for children without stressing any special kind of text that should be given to children on this day.[36]

Talbot's *Reflections* sheds further light on how she values Sunday reading. Published after her death by Carter in 1770, but read by friends before then, *Reflections* is the text in which Talbot most constructively addresses the conflation of work and study in her own life. The text is structured as a series of seven one-to-two-page homilies, each designed to be read on one day of the week. The entries for each weekday describe the pious life by emphasizing patterns of industry and work that culminate in the reader's taking communion on Sunday. In Tuesday's installment, Talbot argues that "constant activity and extensive usefulness is the perception of a spiritual being" (10) and on Wednesday that "Industry makes the world look beautiful around us" (26). In practical terms, the shortness of the pieces is aimed at the working person. That manual laborer Talbot admired and the woman managing a house are invited to see themselves as readers despite their lives being dominated by practical and menial concerns. They are also advised to embrace hard work and avoid closeted study on most days of the week. Like *Low-Life*,

Reflections registers the regularity of modern time, with its increments splicing themselves modestly into a life where duties threaten to take up every hour and where texts need to be read efficiently, in pieces, during minutes stolen from work. But *Reflections* also opens itself up to a different kind of reading on Sunday. It begins with an entry for this day, and it culminates there in anticipation of working readers being able to read more slowly at the week's end. *Reflections* is both a book to be read in its entirety by a reader with leisure to handle it on Sunday *and* a text to be read by the page during the week.

For Norma Clarke, *Reflections* is the text where Talbot finds her stride as a sermonizer, adopting a mode of introspection "licensed by the expectation that elite women had a duty to lead lesser women" (471). This sense of audience is borne out by the popularity of *Reflections*: its thirty-five editions and estimated sale of 25,000 copies suggest that working-class readers found a place for it in the cracks of their lives. But in contrast to the similarly popular *Low-Life*, which spreads itself around the clock but has no time of its own except the continuous present of its own grammar, *Reflections* slants practically and pedagogically toward a reader who will have the possibility to change gears. Without even needing to switch texts, the woman who has no time to read except in snippets on a Tuesday is addressed as the one who will absorb and sit with this whole sermon quite differently if she can give it time.[37] Timing, not page count or genre, determines its length and depth. *Reflections* remains, of course, a highly conservative text. But the way it suggests that a single reader might use books differently at different times has radical implications for the way we think about texts and about readers. It disturbs our ability to pose questions like: What kind of reader was Talbot? What kind of texts did she read? Maybe, as Luhmann might suggest, only reading reads.

Reflections also signals more generally the way other eighteenth-century texts addressing Sunday participate in the organization of time that they describe. Even when texts about Sunday do not explicitly mention reading, they are drawn by their thematic concern with time into the practical question of when the working person is to find time for reading. *Low-Life*, I argued in the introduction, goes some way toward releasing its own reader grammatically from the ceaseless stream of activity that the book represents. It may protest the clash of "church bells and tavern bells [that] keep time with each other across the city" (A2) and scorn the village publicans "beginning to open their Houses for the Reception of early Customers, who had rather arise and go a walking, than dress themselves to go to their Parish-Church" (61). But suggesting that such religious and heathen activities are pitted against each other is a ruse. In fact, *Low-Life* refuses that binary by suggesting a tense for reading that seems not to compete directly with either pleasure-seeking or church-going.

Other texts defending professionals from a 24/7 existence argue for their right to worship on Sunday, to rest and recover at the end of the week, or, in Anna Barbauld's terms, to enjoy the "recurrence of appointed days of rest and leisure" that "divides the weary months of labor and servitude with a separating line of a brighter colour."[38] Eighteenth-century bakers were an important constituency here because they were exempt from Charles II's law prohibiting Sunday labor due to their ovens being crucial in preparing the Sunday meal. They are named as victims in *The Grounds of Complaint against the Practice of Sunday Baking* (1794), a pamphlet that focuses on their right to join everyone else at church. But the pamphlet's urbane style suggests that its real interest is in defending the right of these bakers to become participants of the new public sphere.[39] Decrying an "age of wickedness" in which the Holy Day is no longer respected, a similarly pitched pamphlet reprimands barbers who step in to dress hair on Sundays. Despite what the pamphlet says, the author's outrage seems driven less by Christian morality than by the desire to protect barbers from the pressure to be available at all hours of the week, and to protect Sundays as their own time, time in which they might read and write texts such as this one.[40] Like *Low-Life*, *Reflections*, and Talbot's *Rambler* piece, these essays suggest that when Sunday was being defended from work, it was also being shored up as a new time that might be dedicated to reading. As Altick notes, by the nineteenth century, Sunday was the day on which most workmen read (87).

If Talbot was right, the appeal of Sunday in this regard was not just practical. It also had to do with the fact that the best kinds of reading required a break with the everyday. Picking up books in the hours when one might once have gone to church meant inhabiting a time that had long been understood as distinct from homogenous, linear, or historical time. To borrow the terms of Sunday's defense from the language of religion meant borrowing from a tradition equipped to describe the giving of special time as essential to its own success. The texts I've just described import the temporal logic of religion to the project of education, for which no time slot yet exists. I do not mean to suggest here that the kind of reading that happened on Sunday directly replaced religious activity—although, in another kind of project, that case might be made. It is certainly true that in the later decades of the century, readers felt the switch could be made between reading and going to church. Lady Mary Campbell Coke has no guilt about missing church on a cold Sunday in 1768 because she stays in bed reading the service at home along with one of Atterbury's sermons.[41] The Sunday school movement, which began in the 1780s, suggested that education might practically colonize a time slot once devoted to a second church service. And in the 1790s, Gilbert Wakefield, the radical Unitarian minister, published *An Enquiry into the Expediency and Propriety of Public or Social Worship*, in which he makes the case that Christians might be better off reading for themselves than going to a poor and uninspired service.[42] This generates a new bevy of arguments from writers including Barbauld and Joseph Priestley and Mary Hays for the importance of Sunday worship being collective and institutional, and for the "pleasing distinction in our time" that Sunday introduces.[43] This round of arguments also makes it clear that a certain kind of reading now has the appeal of seeming heartfelt and voluntary in ways that church attendance does not. Going through the motions will no longer be enough. "Insulated discourse," Barbauld argues, must now be "digested into a regular plan of lectures supported by a course of reading" (51). She hopes for public worship and Sabbath-keeping to be rehabilitated by reading, so that Sundays might be diffused with "that air of amenity and sweetness which is the offspring of literature" (59). Ardent readers have become, in short, model users of leisure time, from whom churchgoers might now learn.

But this is not the historical argument I want to follow through. Nor is it the one in which Talbot is most invested as she mounts her case for the pleasures of Sunday reading. What is clear to Talbot, I think, is that just as religion, which she values highly, takes a dedication of time, so does reading of the kind she wants to do more of. The justification for religion's need for time strengthens in her mind the case for book use, not as an alternative to worship, but as one that would require kindred techniques of timetabling. This is the affiliation between the projects, the thing that connects them in her terms as much as content or ideology. It is also what connects fiction and religion in Latour's terms. Both modes are defined, not by certain objects or certain hours associated with them, but as modes that make it known that bringing objects to life takes time, and must be done under concentrated and intermittent conditions.

SOME SUNDAY READERS

It is odd that I have focused on Sundays because they were never really my time of reading. As a child, I was often taken reluctantly on walks or picnics on Sundays. As an adult, I associate Sundays closely with galleries, one of the places I still spend what seems to me like constitutively weekend time. I hear the rain slipping down outside, I relish the bright light on polished floorboards, and I drink expensive coffee afterward among quietly collegial, intergenerational crowds of people in good clothes. Here there is a slowing down in front of pictures one might otherwise pass by: the guard, sitting alone in the corner, a reminder that if one saw the pictures in the context of work one might not even look at them. Even before reading Luhmann, I often thought of this juncture in space and time as a performance in slowing down. You need the pictures, of course, and the walls, and all the cultural capital that gets you there in the first place, but you also need that purposive feeling of not working, of this occasion being different. There is nothing internal to a painting that makes it as visible as the external fact of its appearing in one's vision on a Sunday afternoon.

But my own time for reading was always summer. There were days in my twenties when I lay and read in a hammock every afternoon, though the passage from morning to afternoon would already have become obscure, drinking books in. I am thinking of one summer I spent reading old books on a Danish island, many of them found there on the shelves of the house we stayed in. There was a whole shelf of Ian McEwan, a copy of *Middlemarch* left there by friends who'd brought it from Kenya, a copy of Benedict Anderson's *Imagined Communities* that I'd lugged twice around the world and not yet read, and some Günter Grass in German. Analysis of the books as material sources of transmission could undoubtedly reveal my place in a network of actors, far and near. But nothing, I think now, was quite as material to the way I read them as the shape of that temporal limbo in which they were joined up by *not* being part of my daily life or schedule. The books remain as memorable readings in my mind because I read them deeply, in those warm baggy days away from desk and classroom, much more than because of their other points of contact or association. Was it because I read them in summer that they stand out? I wonder if I remember them so vividly only because that was the last summer we spent on the island without phones and internet?

I turn now to three readers for whom it was Sundays that stood out in their calendar of reading for some of the same vague reasons, as times made by reading, rather than as days given in advance of it. Samuel Johnson, William Temple, and Thomas Turner all admit much more openly than Talbot to being readers. A professional writer, a clergyman, and a businessman, they were all immersed in reading and writing throughout the week, so the question is not whether they read only on that day. The question of *what* they read is also only somewhat inflected by taking the days into account: they all read and wrote sermons during the week, and they all read and wrote secular material sometimes on Sundays. Rather, the question is how a different mode or system of reading co-operates with their sense of Sunday's distinctiveness as an occasion to bring the texts they read then into special focus. Under this lens, books read on Sundays read differently from, say, books read on Tuesdays—and Sundays themselves begin to matter more because of the reading done then.

In beginning with Johnson, I'm choosing the most notorious of all eighteenth-century readers, renowned for his prolific digestion of texts.[44] It seems almost pointless to ask when he read: he must have read incessantly, at all hours of the day. Boswell's *Life of Johnson* is full of images of his spreading books around, picking them up as refuge at dinner parties, and producing text in print-ready form at breakneck speed. But *Life* also documents a string of social engagements so long, and habits of keeping company, not least with Boswell himself, so time consuming, that the question of when Johnson read deeply becomes legitimate. We learn from Boswell that as a child Johnson assumed the habit of taking a book into the fields on Sunday as an alternative to attending church, and thus of reading for two to three hours at a stretch.[45] But this slower kind of reading, detached from his stream of social activity, rarely shows up as part of Boswell's report on Johnson's adult life. The omission points to the fact that Johnson almost certainly read intensively on Sundays, the one day of the week on which Boswell rarely saw him. On Sundays, Boswell tends to dine alone at his lodgings after church and read and write in the evening. Of Johnson, he writes: "He would not have Sunday kept with rigid severity and gloom, but with a gravity and simplicity of behavior" (404), noting that Johnson advises him to observe Sundays (636). When Boswell does interact with Johnson on a Sunday, there are positive signs to suggest that Johnson had been reading and wants to talk about it as a practice. For instance, on Sunday, April 16, 1775, when Boswell visits, Johnson talks about the advantages of reading over conversation (624). On Sunday, May 9, 1779, when they dine together, the topic of discussion is books and literary property (1025). Johnson confirms this as his day of deep reading when he discusses his own ideal program for Sunday activity. "This day, being the beginning of the ecclesiastical year, was a proper time for a new course of life," he writes in his diary one Advent Sunday in 1774: "I began to read the Greek Testament regularly at 160 verses every Sunday" (570–71).

Johnson, who did not attend church regularly, did regularly use Sundays to resolve to do so, as well as to solidify his schemes for the special kinds of reading with which he wanted to mark the day. On Sunday, July 13, 1755, Johnson writes:

> Having lived hitherto in perpetual neglect of public worship and though for some years past not without a habitual reverence for the Sabbath yet without that attention to its religious duties which Christianity requires I will once more form a scheme of life for that day such as alas I have often vainly formed which when my mind is capable of settled practice I hope to follow.
>
> 1. To rise early and in order to [do it] go to sleep early on Sat.
> 2. To use some extraordinary devotion in the morning.

3. To examine the tenor of my life and particularly the last week of it to mark my advances in religion or recession from it.
4. To read the scripture methodically with such helps as are at hand.[46]

Although the apportioning of time might seem to invoke the kind of temporal discipline that Foucault and Thompson associate with the eighteenth century, it is notable here that Johnson is not resolving to read more efficiently or swiftly. On the contrary, he's resolving to read "methodically," by slowing himself down with Greek and with reference books. What lurks within his program of early rising is in fact the resolution to slow down as a reader and to read without the incentives of ease and pleasure, profit and efficiency, that motivated him at other times. By resolving to spend Sundays reading in a foreign language, with a method and at a pace very different from his normal one, Johnson confirms that he saw this day as a day, not just on which to read, but to read differently—in a different mode from the one he generally used.

William Temple (1739–1796), my next Sunday reader, shares qualities with both Johnson and Talbot. A vicar in St. Gluvias, Cornwall, Temple was intensely interested in pursuing a literary life. He met his closest friend, James Boswell, at Cambridge and kept up a correspondence with him throughout the 1760s, 70s, and 80s. Early on in these years, Temple comes across as cheerful enough, happy to serve as Boswell's confessor and confident that his reading equips him better for life than Boswell's excessive writing, which he describes in 1775 as a form of exercising without taking food.[47] Later on, however, Temple bemoans his own failure to read well, writing to Boswell, "I do not wonder now at your reading so little; for when one turns to any author at random, unconnected with any other author, one never engages heartily, one never has time to grow warm" (419). In his published *Diaries*, which begin in 1780, Temple airs even more negatively his worries about his consignment to provincial life, his writing projects, and his plans for a more cosmopolitan life of reading, and Boswell does his best to keep his fears in check and to invite him periodically to London. During these later years of his life, Temple stayed mostly in his parish and worked on his "papers," a proposed collection of essays on "Humour and Politeness, Sensibility, Superstition, Balance on Human Affairs, Parliament, Nobility, Despotism, and Sedition."[48] Like Talbot, Temple was obsessed with making the most of his time and with the difficulty of settling down to use it productively. Lewis Bettany, his early-twentieth-century biographer, describes his diaries severely as proof he "frittered away his time in all sorts of ways and could never husband it in accordance with any prescribed system of study."[49]

Torn between wanting to be part of a more intellectual world in London, and resenting any disruption to his scholarly routines in Cornwall, Temple often writes in his diary when he is depressed by his own failure as a scholar. His young family, for whom he took sole responsibility after the early death of his wife, plagued him: "So large a Family oppresses my spirit and distracts my attention. Difficult to think intensely and watch the flowing of invention with such interruptions and impediments" (50). But he also found travel hard, and when he was away from home he complained in his diary of a lack of routine, poor libraries, and his own social disadvantage. Taking up his pen in London in 1780, he writes: "Find I cannot read or collect here to any purpose, and find myself oppressed for want of my usual exercise. Must never think therefore of reading any where to any advantage but at home, where I can mix study and exercise" (11). In truth, most of Temple's days at home were spent reading or walking. His complaints about time, like Talbot's, are only partly about not having enough hours in the day with books. Just as often, the problem is that his desultory pattern of intellectual pursuit lacks shape and structure: "Still dissatisfied with myself. Must get into a better method. Alas! What have I done to-day. Looked into Boswell's Life of Johnson; rode with Anne. Wrote one letter and read a little in Xenophon, Denina and Cicero. But what is this? Here is no progress, no exertion! Must dedicate the morning to labour, to composition: begin on Monday" (157). Temple chafes here, not long before his death, at wasting time with books rather than writing the one he hoped would bring him immortality.

But in the previous decades he had been more directly concerned with organizing his reading better, and with partitioning it off better from his family life. In the 1780s, this partitioning involved having a new study built at his parsonage, a room where he hoped to spend most of his days. "Try to produce something worthy of notice," he advises himself one Sunday in 1783: "Build your study, buy books, enlarge yr. parlour" (43). While the building of the study went on, with the process and costs of construction and the purchase of books distracting him, Temple fixed his hopes on this possibility of finding a space for his own intellectual pursuits. He was in this sense preoccupied, as Raven has argued many of the middle-class book owners in the period were, with finding a place for his reading and his books.[50] Once the study was finished and occupied, however, his attention turned more directly to the question of how he was to better partition his time. Now he heralds the winter, the season when he is able to hole up without social obligation. Although he still had sermons to write and deliver, Temple is largely free to organize his days as he likes for this larger part of the year, and he several times lays out for himself a new schedule designed to divide his time: "Have been reading in a very desultory manner. Fear I daily grow less assiduous. Emulation damped by distance from Libraries and conversation. Why may I not endeavor to prepare what I have collected. Attempt it on Monday. 9–12 Papers. 12–2 ride. 2–3 dress. 3–4 dine. 4–5 walk. 5–6 read with Nancy. 6–7 Tea. 7–8 read alone. 8–9 Italian or French with Nancy. 9–11 my wife read to me" (86). This schedule was written on a Saturday, and was due, like most of Temple's work plans, to begin on a Monday. But Sundays featured, too, in his reading program, and the fact that he excluded Sunday here indicates just that.

Sundays were days he held apart—not just as the day on which he would preach, but the day on which he'd read differently. At the height of his literary ambitions, when he was determined to train himself better in belles lettres, this meant reserving Sunday as the only day on which he would continue to read sermons: in 1781 he declares he "shall read no more Sermons but with a view to help me in composing. Give up my whole time (except Sundays) to History and Polite Literature" (17). Sundays stand out here as the day he would focus on the study he must do professionally. But in 1790, after a decade in which he has immersed himself in history and belles lettres with disappointing results, Sunday appears again, this time as a way to reinstate a program of serious theological reading from which he has lapsed: "I do hope to be able to read something of what is called Divinity every Sunday, at breakfast and in the evening. Barnard's Divinity of J. Christ demonstrated, against Priestley. Began this to-day but shall reserve it with BP Horsely for Sundays" (88). These two resolutions, made a decade apart, are quite different. One shows Temple hopeful of shaping himself as a literary scholar, but willing to relax his education on Sunday; the other captures him determined to become a more directed reader in the theological canon, and to undertake this course most seriously on Sundays. But in both cases, Sunday is the distinct slot in Temple's self-imposed curriculum. If he is to read fewer sermons, then Sunday is to be the exception to the rule. If he is to read more purposively, then Sunday is to be the day for which the most serious religious texts are to be reserved. As part of a life in which the better use of time was a central theme, both approaches preserve the irregular Sunday slot as one that should not be subsumed to daily routine.[51]

My last reader takes us several rungs down the class ladder, but he also takes us back to Talbot's fantasy of the real worker being suited to become the ideal Sunday reader. Thomas Turner was a grocer and parish councilor in a small Sussex town, East Hoathly. He was also an extremely avid reader. His reading over the eleven years of his diary, from 1754 to 1765, includes a surprising and diverse blend of over seventy texts, ranging from Boyle's lectures, to Cibber, Pope, Sterne, and Shakespeare, Tillotson's sermons, the local paper, the Peerage of England, and numerous texts on history and geography.[52] If the quantity of his reading is in some ways surprising, in others it resonates with comparable cases—for instance, James Woolley, the Nottinghamshire stocking maker whose reading for the early years of the nineteenth century Carolyn Steedman describes carefully in *An Everyday Life of the English Working Class*. Woolley records buying twenty-two books and reading even more, including *Pamela*, *Tom Jones*, and *The Monk*, alongside a range of newspapers and pamphlets. Steedman notes that Woolley presents himself in his diary "as quite at home in the great sea of cultural production in which he swam" (48). But while there is a sense of the library and the hairdresser having been important spaces for Woolley's reading, when that reading happened is less clear.

Turner's reading and his diary keeping, on the other hand, clearly profit from and rely upon the division that Talbot predicts as the workman's advantage: he relishes and enjoys the hours he gets to read in the evenings after long days at work. One Wednesday in April 1758, for instance, he expresses pretty much this sentiment: "Oh what can be a greater pleasure than to be employed in an honest calling all day, and in the even to unbend and relax one's thoughts by endeavoring to improve the mental and more notable part of them" (145). A few weeks later, he reiterates the pleasure after picking up his reading of Collins's *Peerage of England*: "Oh what an unspeakable pleasure it is," he writes, "to be busied in one's trade and at leisure now to unbend one's mind by reading" (146). Part of Turner's appreciation for the evening slot he dedicates to reading comes from the knowledge that he'd spend all his time with books if unchecked in the hours he can give them: "I believe by too eager thirst after knowledge I have oftentimes, to gratify that insatiable humour, been at too great an expense in buying books and spending rather too much time in reading, for it seems to be the only diversion that I have any appetite for. Reading and study (might I be allowed the phrase) would in a manner be both meat and drink to me, were my circumstances but independent" (143). In fact, under these limited conditions Turner often manages to get through more in an evening than Temple records reading in a month. Turner's diary entries casually mention that he's read, or heard read aloud, two Shakespeare plays or a great chunk of *Tristram Shandy* in one sitting (134).

In some ways, Sundays seem to matter less to Turner than his evenings of exposure to books. Sunday is a day on which he rarely, but occasionally, conducts business transactions and usually, but not always, goes to church. He spends many Sundays recovering from nights of heavy drinking, tired and contrite. The plans for a new life that he draws up after church tend to have less to do with reading than with new regimes of abstinence and self-control. The reading he does in this mood is generally quite conventional: Hervey's *Meditations*, Tillotson's sermons, and Young's "Night-Thoughts" are among the things he notes reading on Sunday. On the other hand, Turner, like Talbot, actively appreciates the break in the week's routine, noting "How much more pleasure it is to be home all day of a Sunday and attend the service of the church than to be rioting about as I have too much of late" (254). In his diary entries for Sunday he is more likely than on other days to note a particularly beautiful evening, and it is common for him to praise the quality of preaching he's heard at church. Sundays also stand out as the one day of the week when Temple reads freely during the day, claiming the existence that he imagines men of leisure leading throughout the whole week.[53] It is here that his attention to aesthetics, which seems more finely tuned than normal, comes into play. "Oh what a delightful time it is," writes Turner with unusual flourish one Sunday in May in 1759, "the birds tuning their melodious throats and hymning their creator's praise whilst perhaps man, ah! frail and degenerate man, lies supinely stretched on a bed of luxury and ease" (183). When Turner reads on Sunday, it follows that he generally does so with particular attention to the literary qualities of the text. On Sunday, November 27, 1757, he finishes a round of Tillotson's sermons and deems them "a complete body of divinity, they being wrote in a plain familiar style, but far from what may be deemed low" (125). The next Sunday he reads Sherlock and declares it "a very plain, good book, proper for every Christian to read; that is, rich and poor, men and women, young and old" (126). On Christmas Day that year, also a Sunday, he reads *The Complaint* and comments on the author's "moving and pathetic manner" (127). There are many similar examples among his Sunday entries, none of which would stand out as remarkable were it not for the extremely matter-of-fact manner in which Turner reports on his reading (and eating and socializing) on other days of the week. It seems not to be so much that religious texts rouse his appreciation of style, but that the attention he's inclined to give to texts on Sundays prompts him to note how they work stylistically. The result is that he seems much more attuned to the literary and theatrical merits of texts and sermons on Sundays than he is to their purpose.

Turner's pleasure in reading underscores Talbot's point about the life divided between work and leisure being conducive to a certain kind of happy reader. Turner's diary also allows me to home in on an idea that's run through this chapter so far, which is that a mode of reading was emerging in the eighteenth century that we would now call literary, or even critical, in which the dedication of time to a book was crucial. For Talbot, the point was that only compartmentalization of time could allow her the freedom to read as she wanted. But we've also seen the experience of time being variegated by the willingness of readers to calibrate the speed and quality of their reading to the different days of the week and seasons of the year. The causality remains complex, with reading providing the break with everyday life on which its execution also depends. But it is in this conjunction that the truth of book reading as a fulfilling kind of time use becomes clear in Talbot's terms. Any model that took reading's occurrence in time as something more empirical or as a simple kind of coordinate would miss, I think, this performative contribution to the way time given to books was felt as something deliberate and distinct from daily routine.

SIR CHARLES COMES AND GOES

I have not focused very much in this chapter on what kind of literature was being read in these intermittent bursts because I wanted to push past that question of content and focus instead on how modes of reading might provide their own temporal zoning. But I turn now to a text we know Talbot read in manuscript form to her satisfaction: Samuel Richardson's *The History of Sir Charles Grandison* (1753). Schellenberg writes of Talbot spending "eighteen months or more" working with the long manuscript in an intensive editorial role in 1751 and 1752, reporting that she's reviewing it "as Carefully and I could almost say as Conscientiously as I can" (86). But the fact that these are years in which Talbot was also complaining to Carter of the continual stream of household duties that kept her from study raises the question of how the *Grandison* manuscript registered for her. Was editing her friend's manuscript easier to accommodate as concentrated work than just reading, or did *Grandison* itself appeal to her in some way as receptive to the intermittent but focused attention she wanted to give books? In the case of *Grandison,* the practice of reading a book in parts, and at certain times, resonates with Sir Charles as a character whose presence is spread out unevenly across the regular time of the novel. This means, I'll suggest, that one form of temporality associated with the reception of the novel is at odds with the homogenous-continuous form of time suggested by its long catalog of letters. Richardson's meticulous situating of fictional events in real time, so crucial to his role as innovator of realism, is not the only work he does in representing time. Reading, I want to argue, becomes part of his plot as an antidote to the excessive continuity and presence of realist narrative—and Sir Charles himself becomes aligned with the selective use of time that comes into being as the book is handled, rather than with the round-the-clock life upon which the text seems to report.

When Richardson was composing *Grandison,* his third and final novel, he relied on input from several members of his circle of correspondents and friends.[54] To Talbot's mind this meant its lead character was in part her creation. She wrote to Carter: "Did you ever call Pigmalian a fool, for making an image and falling in love with it—and do you know that you and I are two Pigmalionesses? Did not Mr. Richardson ask us for some traits of his good man's character? And did we not give him some? And has he not gone and put these and his own charming ideas into a book and formed a Sir Charles Grandison?" (1:291).

There are, of course, many ways in which Sir Charles, embodiment of male openness, sincerity, and good judgment, might have appealed to Talbot. But, as someone who worried so actively about temporal economy, the novel itself seems unlikely to have pleased her. The letters that make up its thousand-odd pages consist of letters written and circulated by a set of well-heeled spendthrifts of time over a period of less than a year. The members of the Grandison household—made up for most of the novel by Harriet Byron, letter writer extraordinaire, Sir Charles's two sisters, and his teenage ward, Emily—spend their days producing and reading letters, many of which do little more than extol Sir Charles's virtues. All the time that might have been spent in the kind of active duty that Talbot describes in *Reflections* is eaten up with writing. "I have written," declares Harriet, "for these two days passed at every opportunity, and, for two nights, hardly knowing what sleepiness was, two hours, each night, have contented me."[55]

While the most obviously identifiable theme of the novel is that of elective marriage, the narrative builds up this concern by situating its characters in an environment where multiple partnerships and affinities seem possible at all times. As several critics have noted, the novel overflows, not so much with action, but with lines of possible attraction, all of which must be brought to life through letters. Sir Charles, who incites feelings in every woman he meets in England, is also, for instance, central to another set of effusive relationships in Italy, where he has been all but engaged to Clementina. As fast as he reads letters, executes wills, and brokers marriages other than his own, new fountains of discourse, money, and adoration seem to spring up. Leah Price has described *Grandison* aptly as a novel in which "multiplicity leads to surplus and surfeit. The text is too long, letters circulate to too many readers, the hero has too many virtues, the characters have too much money, too many characters are in love with Grandison—and, if only by a margin of one, Grandison loves too many of them in return."[56] Indeed, if ever there was a text to make a reader feel oversupplied with pages, and undersupplied with hours in which to read them, it seems to be this one.

This sense of lack might have resonated with Talbot, who felt that she was the victim of it in her own life. But it also helps suggest why she approved of, and in fact felt partly responsible for, Richardson's solution to the problem he was exacerbating as a novelist. To make Sir Charles a figure who supplies the forms of division and distinction that the plot lacks is to intuit the way in which the book will have to be parceled up as readings across time. *Grandison* is, Wendy Jones argues, a novel in which "making distinctions itself becomes a primary focus."[57] In a world where readers and women more generally seem bound to feel overwhelmed by competing demands on their time, Sir Charles knows how to manage things. Presented by Harriet as a man whose "time, you see, is very precious," Sir Charles also identifies himself as busy. "When," he asks rhetorically in one letter, "shall I find time for myself?" (iii:ii). Yet, Sir Charles never forgoes sleep in order to complete his letters. In fact, he proves expert at protecting his hours. Typically, we'll find him asking when dinner will be ready in order to arrange a window of time for a discussion, or extracting himself elegantly from neighborly visits in order to finish a business affair (vi.i). His departures and arrivals happen at his bidding, when all are asleep or have ceased to expect his arrival, and his private audiences are called for at times convenient to him. Despite their importance, and the efficiency with which they circulate, Sir Charles's own letters are also written with relative economy and often disseminated by his various "sisters" on his behalf. He controls their pacing by using the time it takes for them to arrive as a resource to his advantage. And his reading of others' letters often happens in hours of allocated private withdrawal, for which he makes time.[58] In terms of the novel's plot, this command over the fort/da of his own presence helps preserve Sir Charles's credibility in what would otherwise be the riotous plurality of his affections, his "divided or double love" for Harriet and Clementina (v:xxviii).

It is worth thinking here about a case Latour makes in "On Interobjectivity," an article published the same year as his conversation with Serres, which I discussed in the introduction. Latour's ostensible purpose in this argument is to define the distinction between human and simian interactions. After stressing the similar levels of complexity with which baboons and humans interact with each other and the material world, he arrives at the conclusion that what distinguishes us as humans in the end is only the way we are able to cordon off our interactions. A counter at the post office, to use Latour's example, creates the space of a dedicated interaction—however complicated the sale of stamps gets, what goes on there remains fenced off from the complexity of general existence. We won't get caught there talking about our work, or our health.[59] Sir Charles is a master of such cordoning off and, as Latour's model suggests, he perceives how much depends on his keeping his affairs distant in space. In his presence, doors and rooms become, for instance, barriers in ways they rarely seem to be to Richardson's female characters, who are notoriously being burst in upon in even their inner sancturns. That Clementina lives in Bologna is absolutely vital to the decorum with which he manages to love both her and Harriet, someone with whom he regularly shares a house in England.

Yet Serres's 1990s rejoinder to Latour is also one from which the literary critic might learn. Tempting as it is to read Sir Charles as creating divisions in space that serve his mastery of affairs, we should not forget that he is just as active in creating divisions in time. In his initial negotiations with Clementina, for instance, patterns of absence and presence determine his vision of the marriage that might be. This involves his proposing at one point a schedule of alternate residency in Italy and England that the couple might undertake, and at another point conducting a courtship in which timing is crucial: "I will," announces Sir Charles, "absent myself for some time from Bologna; but (as she has the goodness to acknowledge an esteem for me) with her leave. I will return at my time. I will repeat my absence, if we have the least shadow of doubt" (vi.xii). Space comes into play here, but timing is just as important: Sir Charles's ability to make himself appear and disappear sequentially, to make his own presence intermittent, matters more than his ability to shut the door firmly on his study. If, for Latour, the work of being human depends on shutting out the background noise through walls and screens, and focusing, say, on one text for long hours at a stretch before going off to cook or argue, for Serres it involves a process much more combinatorial. "The living organism," he argues, "is of all times. This does not mean that it is eternal but rather that it is an original complex, woven out of all the different times that our intellect subjects to analysis or that our habits distinguish or that our spatial environment tolerates."[60] In these terms, what matters is not just that Sir Charles can come and go, but that he appears able to move at his own pace, even as Richardson regulates his appearance as a character.

Talbot, I've been arguing throughout this chapter, wishes to be an organism of this kind. Sir Charles already is. But the women in *Grandison* are not. As characters and writers, they are tied to the mast of clock time, from which there is no retreat. Just as their closets are frequently broken into, so are their days porous to events, unplanned arrivals, and the needs of others. Their letters are written to the clock, even if they manage to extend the hours they have alone by providing an incentive for going without sleep. They live, in this sense, like Talbot, or like the primates Latour describes, besieged at all moments by the convergence of different spheres of their lives they cannot isolate from each other. Their only response is to exercise control along this single axis of time. Given the job of naming her wedding day, Harriet thinks about the timing of her romance with Sir Charles as one everyone has shared, mentally consulting a common calendar before giving her answer (vi.xxix). And when Harriet and Emily feature as ideal household managers, it is because they are conscious of the single flow of time that unites everyone. Harriet is praised for maintaining "a succession of orderliness" in the house: "One right thing is an introduction of another; and all is in such a method, that it seems impossible for the meanest servants to mistake their duty. Such harmony, such observance, yet such pleasure in every countenance!" (v:xvii). As the result of her regime, "the servants have generally time to themselves, an hour or two in a day" (v:xvii). But Harriet does not get to extract herself in this way from the action. The time she has alone, she spends writing the letters that lock each day's events more efficiently into diurnal time. Her being is ranged along the line of one single temporality rather than combining many.

Harriet's time management therefore entails a very different project from Sir Charles's, which remains to the last reliant upon temporal sequestration. His response to his own betrothal is to absent himself from the story until Harriet has fixed its timeline (vi.xxix). Like Talbot's romanticized workman, Sir Charles exhibits an ability to move concertedly between multiple temporalities, to dip in and out of different moments without feeling the need to link up or anticipate events. His and Harriet's strategies might be compared here to the two modes of narration that Stuart Sherman discusses as the older and newer influences on the eighteenth-century novel. For Sherman, novels like Richardson's, with events clocked as passing days, inherit the logic of the diary and the periodical: they model "calibrated continuity as a paradigm for possessing time as a property and managing it well."[61] This involves Harriet's mode of accounting for everything that happens as it occurs and for her creation of the "succession of orderliness" from which there seems no break.

The older mode, the one that resists diurnal form, is more critically selective, and it involves taking elements for description and piecing them together discriminately, without any pretense that the narrative follows the natural course of time. This is exemplified for Sherman by Samuel Johnson's prose style. But it also has its equivalent in Grandison's comings and goings, his appointment settings and deliveries of news, all of which are episodic in their movement, disassociative rather than connective in their representation of experience. If Harriet's behavior conforms to the logic of codex as a linear movement through time, Grandison's suggests the importance of the chapter, a form, Nicholas Dames points out, that "openly permitted a reading oriented around pauses—for reflection or rumination, perhaps, but also for refreshment or diversion."[62] As Sherman shows, it is Harriet's view of time that is in ascendency when it comes to narration. But it is Sir Charles's approach to time that resonates much more strongly with the way many eighteenth-century readers were reading (and not reading) books—especially ones as long as *Grandison* itself—as chapters and as intermittently present objects.

At one point Sir Charles is described as being like a book. But we might think of him as less like a book per se than like a read book—a book as it is taken up and opened, say, on Sundays. His own ideal form of attention and presence, calibrated as it is to accommodate a world of multiple temporalities, resembles the art of reading much more than that of the chronologically arranged text. Sir Charles's character also recalls Derrida's understanding of the gift, which exists in "a delimited time, neither an instant nor an infinite time, but a time determined by a term, in other words, a rhythm, a cadence."[63] If Harriet writes a modern narrative, Sir Charles stars in it as the modern reader—one short of time, but able to dedicate himself to reading in bursts of intensity and concentration as long as he is not required to stay all the way through—and as long as he can close the covers on a scene and attend to other kinds of work when he wants to.

This might help explain why Talbot, despite all indications to the contrary, liked *Grandison* so much as a novel, and owned Sir Charles so readily as her own creation. It is Harriet, to be sure, who models Christian efficiency in her work. But it is the logic of temporal differentiation that Sir Charles brings to a world overcrowded with events that Talbot admires more. It is also the logic by which she accessed Richardson's novel even when short of time, reading it closely in manuscript form over a long period, and then referring Carter enthusiastically to particular scenes for deeper study.[64] At no point in her advocacy of *Grandison* does Talbot suggest that she's read it from beginning to end, or that she's captive to its sequence. That feeling she reserved for the way she felt about her actual days, the ones whose minutes seemed to run too fast through her hands, and from which there was no relief. Sir Charles's mode of being there sometimes and not at other times is thus ideal to Talbot for many reasons, but above all it is ideal to her as a model of the relief good reading should provide from that single temporal economy.

Among those who now share Crary's objection to smart devices and lives led online are many who advocate the book as a better platform for reading, a platform that was designed to promote division and concentration, to sequester the reader from a culture of constant communication, and one to which we might return—or at least teach our children to return—if we want to unplug.[65] But we've now seen many eighteenth-century readers struggling with many of the same issues Crary worries about. Harriet and her sleepless nights catching up on the things that happened each day; Talbot and her broken, wasted days; Temple and his search for that elusive place of study: they all already lived in worlds where they felt there was no escape from schedules, duties, and paperwork.

New kinds of print could be seen as part of that problem. But books, which could seem by dint of their length and quality to incite another kind of attention, were not machines that automatically allowed one to switch off. As Natalie Phillips shows in *Distraction*, attention has always been an issue for readers: one could glance at a book or read it superficially. Good reading was a matter of giving a certain kind of time to an object, not a function of its materiality. It was like going to church or the gallery, visiting in loops of return and in a certain mood a site that was always there, but could not always command devoted attention. In return, what one got from it was the protection of temporal variation. The downside of this argument is that it suggests books are not the answer. There's no book that will unplug you or force you to concentrate. The upside is that it makes the attention we give texts something that's always been conjugated in time and has always supported the idea of temporal differentiation. This might help us to understand in a light other than the one Crary sheds on media why fiction as a mode and art as a system have always required a slowing down and a tuning out of everyday life.

CHAPTER 2

Joining Up Time

> How we waste our labours in reading without system—or even with system. I have been led to think again of this by looking over some notes of reading that I took years ago, and finding I had forgotten them entirely. To stick to a few books and read these over and over again is the only way.
>
> —Thomas Hardy to Florence Henniker, September 11, 1895

REREADING FOR HAPPINESS

That copy of *Wuthering Heights*: unscholarly but well bound, a little puffy and good to hold. The font put me off for a while at fifteen, but I settled down and read it eventually because it was a gift from someone I wanted to impress. I don't remember the reading except that I was not a convert; all seemed dark. But I do remember the book itself, its pages intermingling with my own teenage sadness and longing. I read *Wuthering Heights* again in college, choosing it as an optional text for a course because it was one of the few books I owned. The old copy came out again in its blue leather cover, absorbing smoke and wine in the evening in the Brighton flat I shared with friends. This time I remember the reading better, and the essay I wrote back then about gender relations in the novel, which I was now inclined to critique. The same dark text I'd read in my unhappiness now mingled with my exuberance about literary theory. More recently, I read it a third time after agreeing to teach a course on nineteenth-century literature. *Wuthering Heights*: Did I know it at all? Did I even have time to read it properly? Read during term time, it sank in slowly, but eventually I wrote a lecture and finally an article about the plot, still unsure about whether I liked it. I used a scholarly edition but peeked back at that old one, still on my shelf, the one that's inscribed "To Tina, with love, Christmas 1986."

In *The Value of the Humanities*, Helen Small surveys briefly the way literature has figured in recent cognitive studies of happiness. She finds us ill equipped to say why an institutional familiarity with books should help us live better. Sociologically, there's not much to be said. Most of us who read regularly belong to a group who would have a range of good things to keep us occupied even if we lacked books. But in navigating her way back through the debate, Small suggests there might be some value in revisiting John Stuart Mill's account of happiness. His utilitarianism states that the bar of happiness is raised by the elevation of understanding, both individual and collective, that comes from an engagement with the arts. Small boosts this theoretical point by applying it to Mill's *Autobiography*, where literature features at moments when Mill grasps changes to his life *as a life*, sensing its longer, unstable course. While his social theory offers a decent justification for the arts, it is Mill's own story of familiarity with books over a lifetime that provides the real case for the importance of reading to happiness. The reader able to draw on a reservoir of literary knowledge has access to a special structure of narration when it comes to "the temporal and often uneven development of an intellectual and ethical life."[1] Here, ironically, the point is not about books as a source of pleasure or continuity, but about the way they allow one to parse discontinuity. Conjuring up books to explain moments of change, Mill relies on them as tools in hinging the disparate parts of his life together.

The last chapter explored the temporal spacing out of systems that enabled readers to tune into texts at certain times against the background noise of daily work and distraction. Here I am interested in the lifetime, rather than any single day or season, being the time in which we read. In looking at three different readers—Elizabeth Carter (1717–1806), William Grenville (1759–1834), and Amelia Opie (1769–1853)—my focus is on the way they revisit and rework their understanding of texts. If books served in the last chapter as things to be picked up and parceled out at intervals of forgetting, in this one they appear as touchstones in a practice of revising, of packaging up our differences as story. Reworking one's position as reader of a certain text becomes a way of recouping time as life and discontinuity as critical maturity. It is in this interwoven fabric of readings spaced out over time, I want to argue, that a certain kind of happiness is to be found. It is because they differ from each other that my readings of *Wuthering Heights* pile up. No one exuberant reading would suffice to explain why my relation to this novel makes me happy.

The rereaders I'll be describing in these terms are not of the kind most common in the eighteenth century. The most frequent rereadings of the period were obviously of religious texts. But I'm not so concerned here with the weekly user of the Bible, or the Sunday devotee of books whose habits were in focus in the last chapter. This is not because reading a spiritual text at different points in time can't contribute to a narrative about life as change—as conversion narrative, it obviously does just this—but devotional reading just as often produces other chronologies of growing familiarity or appreciation of a text. As a habit, as a path of return, regular and familiar reading of a text whose authority one trusts is easily justified. It differs in this respect from the style of return this chapter describes. It's not just that Carter's rereadings of Epictetus, Grenville's of Aristotle, and Opie's of William Godwin are critical; it's also that they must produce their own form of justification for lives spent in study—one akin to the case Small is making for the role of reading in Mill's life and in the humanities in general.

For the same reasons, I'm bracketing out here the kind of rereading that Deidre Lynch describes as a "going steady" with books. Lynch's *Loving Literature* generally does the important work of documenting the late-eighteenth-century emergence of "an idea of literature as that which we are always reading and never reading for the first time."[2] It helps correct in this respect the tendency of many theorists to invoke a model of reading where the reader is always encountering a book for the first time. I'll be following Lynch's footsteps in this regard in my critical engagement with Vivasvan Soni's *Mourning Happiness*. But Lynch's focus in this part of her study is literature as a site of devotion rather than reading as a practice through which complexity, critical acumen, or understanding might build up. In general, *Loving Literature* is concerned with readers whose approach to literature almost aggressively rules out such critical trajectories. Lynch points to fans of Austen and Wordsworth who revisit their work on an annual basis, handling well-known pages with a deep investment in the steadying influence of works they know and admire. Between the 1760s and the 1820s, she argues, a culture of reading arose that could "accord a surprising amount of respect to affections of a more everyday, more placid and even torpid cast, affections directed at texts so deeply familiar that the emotions that they stir barely register at the level of consciousness."[3] Books in this account mirror, soothe, unite. Well into the nineteenth century—well into the present, even we, the new lovers of literature might want to admit in the face of Lynch's persuasive argument—the almost drug-like quality of rereading the canon features alongside the history of books guiding us toward new ideas.

This torpor of affection is important, but it is not what Small has in mind when she thinks of Mill and his knowledge of a literary canon as the basis for his *Autobiography*. And it is not what I have in mind in my relation to *Wuthering Heights* (though I might have told other stories: my daughter, never without a copy of *The Song of Achilles* in her overnight bag, an elixir to soothe her to sleep; my son, telling himself *The Lord of the Rings* over and over again as he plays in his room).

The rereadings I am interested in accrue. They build up, not as a deepening relation to or belief in a text (I do not necessarily know *Wuthering Heights* better now), but as a changing form of involvement with books that justifies returning to them in the spirit of professional occupation and achievement.

Approaching rereading as a process of development and change involves looking at a specific relationship between a certain kind of reader and her text over a longer stretch of time. But it also involves another, more pragmatic shift in the way we do the history of reading. While it would be perfectly legitimate in historical terms to find evidence of Grenville's opinion on Aristotle in his student days, or Opie's rejection of Godwin's political writings in her middle age, in this chapter I assume that such readings are subject to revision. Just as loving literature involves a pattern of return to a text, so does critical reading, if it is concrete at all, involve a path of return that becomes visible when more than one event is involved. Even without Serres's and Luhmann's cases for this being the proper way to study any network or system, common sense as literary professionals can tell us this: our critical readings of books depend on our revisiting them in unstable ways over time as much as our uncritical readings do. Heartfelt and pious relationships to books do not have a monopoly on reading as a practice of return.

Ultimately, then, there are two very different concepts of the good life, of happiness, nestled within that question of how we spend time with the books we know: one that is connected to pleasure, agreement, and familiarity, and another, just as important, connected to change and the different concept of happiness that comes with that. It is summer now. Facebook seems flooded with stories of academic friends rediscovering Proust and James and unfashionable critics, long forgotten. We seem to like this kind of rereading on every scale as something that keeps us alive. I have given my daughter that old copy of *Wuthering Heights*. She doesn't like it. I tell her: "Fine, just read it anyway so that you can remember not liking it when you read it again." This argument does not convince her but I am pretty sure it's a good one.

SLOW TRANSLATION

Elizabeth Carter's edition of *All the Works of Epictetus* (1758) is an impressive thing. The folio volume consists of four of Epictetus's previously untranslated dialogues and a new translation of his potted advice on the good life, the *Enchiridion*, popular with eighteenth-century readers as a self-help manual. Carter's edition moves Epictetus away from this audience. It is unabashedly the work of a scholar. Her introduction carefully weighs the work of the Stoics, considering particular difficulties of translating from Greek and annotating Epictetus's ideas for an open-minded, modern reader. Carter herself is in evidence, not only as translator, but as a general guide to a Stoic text eighteenth-century Christians were likely to find troubling. The translation appeared with a long subscription list of powerful names in a handsome edition that was well received by the public.[4] Its sale funded Carter's purchase of a house in Deal on the coast in Kent and a life of modest scholarly independence. Preparation of this book was central to her reputation as a scholar and, I'll argue here, to her conception of the good life. But *All the Works of Epictetus* remains Carter's only significant publication. Apart from a handful of poems and short essays, and two less significant translations, Carter demurred from publishing, or indeed from writing more than letters, in her lifetime.[5]

This limited literary output has made it possible to portray her as an author whose fate might have been different had it been she, and not the younger brother she educated, who went to Cambridge; had it been she, rather than her friend Samuel Johnson, who set the tone for mid-century literary exchange. There have been several readings of her life and writings motivated by the desire to recover Carter as a writer, notably by Elizabeth Eger and Judith Hawley.[6] But Carter plays a different role in this chapter, where I want to draw her out as a prolific, multilingual reader whose accomplishments on this score were already greater than those of most men around her, perhaps largely because of her unwillingness to write more than she did. Carter also features in contrast to Talbot, whom we met in the last chapter as a reader short of time. Carter views study differently, not necessarily because she does more of it than Talbot, but because she represents her access to books over time differently. Carter's reading involves a robust way of spending time that is scholarly, critical, and, in her own account, often successful. This lies less in the hours she clocks with books and more, I want to show, in the way her practice of reading, and rereading, allows the fabric of her life as a *lifetime* to emerge.

It is key here that Carter, in contrast to Talbot, had a relatively high level of control over her own daily activities. She was protected in her autonomy by living in Kent, by a clergyman father who valued her independence and intellectual accomplishment, and by the relative affluence of her friends. Elizabeth Montagu, who eventually settled a pension on Carter, complained of "an appetite for reading which my modes of life hinder me from satisfying."[7] Talbot, as we have seen, labored her whole life with the feeling that social obligations in London and charitable work in her parish left her no time for study. But Carter, further down the social ladder than either of these contemporaries, pursued and was recognized during her long life for her more fully realized relation to books. The letters between Talbot and Carter reveal her busy with reading, translating, and exercising in a combination that often leaves her content.

Much of what we know about Carter's life comes from her nephew, the Reverend Montagu Pennington, who published the hagiographic *Memoirs of the Life of Mrs. Elizabeth Carter* in 1825. Here he describes his aunt's impressive skills as an active reader of Greek, Latin, Hebrew, German, French, and Italian, languages she acquired as a young scholar. He praises her for having been so determined to read that she rose in her youth before dawn and experimented in chewing snuff and drinking coffee and green tea to stay awake over her books at night.[8] Pennington comments with surprise on her keeping up such a regimen even once she is known for her learning and "courted by all the principal families in the eastern part of the country" (100). "Her general rule," he states in his spirit of austere admiration, "was to read before breakfast two chapters in the Bible, a sermon (among which she gave preference to Clarke's, Secker's, some of Sherlock's and all those of the late Bishop of London) some Hebrew, Greek, and Latin. After breakfast she read some part of every language with which she was acquainted, so that she never allowed herself to forget what she had once known" (140).

Carter's own account of her days at Deal, given in detail at Talbot's request, describes this regimen more lightheartedly by suggesting that her study is interspersed with distraction and social engagement: "Between reading, working, writing, twirling the globes, and running up and down the stairs an hundred times to see where everybody is, and how they do, which furnishes me with little intervals of talk I seldom want either business or entertainment."[9] In general, Carter's correspondence gives the impression that she usually managed to devote many hours a day to books, reading things of note as they were published, and regularly revisiting the classics she valued. Seen from this angle, Carter's problem was less one of finding time to read than of representing the time she spent with books as worthwhile.

This challenge was further complicated by the fact that Carter's reading, while wide-ranging and often up-to-date, was directed toward reading and rereading texts in ancient and European languages. She is so ensconced in these languages that she complains in a 1754 letter to Talbot of having lost her taste for English: "Do not be angry with me if I confess, that I am lately grown a little out of humour with English, to which I used to be so partial, but translating out of Greek has helped me to discover some very provoking defects in it, which never did me any harm before. To complete its downfall from the elevation to which my prejudice had raised it, my walking companion for the last half year has usually been some Italian poet; and of Italian I am grown fonder than ever" (2: 175).

Reading so well in languages other than English obviously fueled Carter's pleasure in sustained textual scrutiny. In 1745, reading Dante anew in Italian,

Talbot writes to Carter about how pleasurable she finds old books, which "bear reading very often, and I think there is full as much pleasure in reading a very excellent book the fifth or sixth time, as if one had it fresh from the press" (1:101). To this, Carter responds enthusiastically that she too most enjoys "the fifth or sixth" reading of a good text (1:106). Carter rereads essays over which she and Talbot disagree, texts in languages she's keen to retain, texts she admires, novels she doesn't like at first, and the classics of which she's a scholar. But her reading's main trajectory involves secular and interpretive loops of return—especially, of course, to Epictetus, the author she translated professionally.

Perhaps this is the way many of us read—though it's hard to track; in fact, it's hard to differentiate between the way we go back to certain texts as talisman and others, like newspapers or websites, for the combination of novelty and a form we know, and still others as things we are always working with. Certainly, it's not generally the way we imagine eighteenth-century women, who are often seen as craving news and plot, or as thirsty for the latest items from the circulating library. Carter stands out from this imagined crowd of readers as having access to, and encouragement to use, her father's library of classical literature, but she's not the only one in this position. In other such cases, however, rereading can be harder to track.What makes it uniquely possible to observe patterns of rereading in Carter's case is the paper trail that her process of translation left.

Her translation of Epictetus began and remained for almost two decades part of a back-and-forth with friends. In 1743, Talbot mentions in a letter to Carter that Epictetus is one of her favorite authors to read in the morning. Unable to read Greek, and impressed by her new friend's linguistic abilities, Talbot declares herself "infinitely provoked" that there is no translation of the parts of his precepts beyond the *Enchridion* (1:42). In June 1749, Carter begins in response to send Talbot "scraps" of her own translation—proof, as she puts it, of "how idle I have been" (1:313). She finds a willing recipient in Talbot, who, together with her mother, welcomes these "scraps" and begins to copy them into a "little book" they keep for this purpose: "I admire," writes Talbot, "Epictetus more and more every day . . . there is a nobleness in its simplicity very striking. A superiority of thought and shortness of expression that makes both my mother and me wish for more" (1:317). By 1750, Talbot's friend Secker has become part of this circle and there's a steady stream of translation flowing through his office en route between Carter and Talbot. It became their standard practice for Talbot to submit Carter's translated pages for comment to Secker, copying his notes along with the translation into her little book, and then sending the comments and any thoughts she had back to Carter. A letter from March 1750 begins: "After a tedious time, dear Miss Carter, I return your Epictetus with my Lord's remarks. I have had them this fortnight, but really have not had time to copy them till to-day" (1:328).

Between 1751 and 1755, the translation takes more of Carter's time. Even when she describes herself seized by "a most violent disinclination to writing" she encloses in a letter to Talbot "a deal of Epictetus" and promises that more is ready (2:29–30). This suggests that translation figured positively in her terms as reading, rather than as the writing she is not doing. Yet at periods during the next years, Carter loses faith in the whole process of preparing her edition, confessing to Talbot and Secker the efforts it takes to buoy herself up in completing it: "As to any lions or bears which my own imagination may have conveyed up against this undertaking, I am determined, most heroically, to knock them all on the head." (2:142). By 1755, her taste for the process is strong again. When the draft is finished, it continues to traffic between London and Kent, with Secker locking himself up with it for a month in order to produce a last round of comments, and Talbot correcting and introducing its final form, preparing critical notes and an introduction for the edition. In 1760, when the translation is out in the world, Talbot is still commenting to Carter on the pleasure she takes in this "treasure" supplied by her friend (2:309).

Translating and reading were, then, connected as part of a dialogue and exchange in Carter's case. The decade-long process of her translation, which spilled into further years of friends reading and weighing in on *All the Works of Epictetus*, appears for much of this time as constructively argumentative as well as progressive. Talbot, for instance, urges Carter on with the translation, but also stresses that she expects Carter's notes to "mark out those points in which [Epictetus] is false, wild, and defective, and to draw comparisons between that, and the only true philosophy, the Christian" (2:138–39). In this Carter obliges, lingering over the preparation of the edition, using the scholarly footnotes to respond to Talbot's concerns, and awaiting Talbot and Secker's thoughts on these notes. At one point, for instance, Carter's edition pauses to consider why Epictetus dwells so often on "externals": "Readers, perhaps, may grow tired, with being told so often what they will find it very difficult to believe, That, because Externals are not in our own Power, they are nothing to us" (373). We can imagine Talbot here, in her spirit of frustration, as the person Carter addresses. The language of Carter's footnotes, which refer often to the different "readings" passages might have been given, is also a reminder that Carter's translated pages of Epictetus traveled between the desks of Talbot and Secker as *readings*, texts that kept all three actively revisiting and criticizing a writer with whom they were by now all deeply familiar.

Carter's translation can be read in this sense as evidence for her preference for books that reward interpretive relationships extended over time, and for a practice of reading closely intertwined with one of writing and dialogue. Michael Warner argues persuasively for critical reading as "largely projected from our own circulatory practices . . . it is indeed an essential element of critical reading that the reader be imagined as a producer of discourse."[10] But if Carter can be cast in this light as a critical reader, in thrall to the pleasures of scholarly interaction and writing, she can also be understood as one drawn to translation as a form of labor that renders reading compositional. Clive Scott's *Literary Translation and the Rediscovery of Reading* urges this view of translation as comment on the way all books are used over time. "It is," Scott observes, "the translator's business to put the ST [source text] at the cutting edge of its own progress through time, to open up for the ST its possible futures, its strategy of self-regeneration."[11] The greater the richness of the original text—its translatability in Benjamin's terms—the greater the plurality of different readings that constitute its future.[12] Thus, while reading the same text can self-evidently be a form of repetition—a "going steady" in Lynch's terms, a reading on the spot—once it is enacted as translation, more emphasis falls on the differences that come with rereading an original text over time. This idea of translation makes Carter's biography as a reader distinct from ones focused on the acquisition of erudition. Carter's translation justifies her progress as a rereader in terms quite different from those deployed, for instance, by James Lackington, who enumerates the books he has read and written. Carter, who regularly rejects invitations to write more, and shows absolutely no envy, for instance, of her brother's formal course of education, seems determined to have her reading leave a different kind of trace. As someone who clears hours of her day for reading but shuns much that is new and acquisitive, she makes reading a limited portfolio of texts into something that shows up as time use without being temporally progressive. Measured in terms of her other outward accomplishments, Carter's reading bears little obvious fruit: it takes time but is its own end.

It is telling that Carter translated Epictetus, a direct opponent of thinking in teleological or instrumental terms. His discourses consist of a witty but unerring stream of advice regarding self-mastery and freedom from desire. The ode by Hester Mulso Chapone, which appears on the title page of Carter's translation, celebrates Epictetus in these terms:

No longer let my fleeting Joys depend
On social, or domestic Ties!
Superior let my spirit rise,
Nor in the gentle counsel of a Friend,

Nor in the Smiles of Love, expect Delight:
But teach me in *myself* to find
Whate'er can please or fill my Mind.

But while his philosophy clearly concerns mental development and autonomy under conditions of external restraint, themes close to the hearts of those in the Bluestocking Circle, his dialogues push back against the idea of anyone working to educate themselves as a way to move up or outwards in the world. The stoic activity of perfecting one's soul scorns outward measures of accomplishment, invoking instead the happiness to be found in achieving freedom from one's desire for such ends. Asked about progress, Epictetus replies by imagining the reader who says proudly: "Take the treatise on *the active Powers* and see how thoroughly I have pursued it." Epictetus rebukes the reader by stating that this is no way to measure bookish achievement: "I do not enquire into this Wretch, but into how you exert those Powers; how you manage your Desires and Aversions, how your Intentions and Purposes; how you are prepared for Events, whether comfortably or contrary to Nature. If comfortably, give me evidence of that, and I will say you Improve; if contrary, go your way and not only comment on these Treatises, but write such yourself; and what Service will it do you? Do you not know that the whole Volume is sold for half a Crown?" (16). It is easy to read this rebuke as Carter herself speaking, and speaking quite directly to Talbot, of the idea that reading is only as good as its reader, only valuable as an activity if it answers its own ends. Later Epictetus dismisses reading books on these grounds as an activity to which other forms of life should be sacrificed. It is Talbot who seems to speak when his interlocutor exclaims: "I am in a wretched Way, I have no leisure to read!" To this, Epictetus replies: "For what purpose would you read? Tell me. For if you rest merely in being amused, and learning something, you are insignificant and miserable. But if you refer it to what you ought, what is that but a prosperous Life? And if Reading does not procure you a prosperous Life, of what Use is it?" (372). When the unsatisfied reader explains that it is because reading secures prosperity that he is uneasy at being deprived of it, Epictetus asserts that such a feeling is at odds with pleasure in reading as an end in itself. The only correct motivation for reading is pleasure in the posture of a good reader; the only proof of the reader's achievement is that she reads well.

There was much that an eighteenth-century Christian reader of the Stoics could object to, and Carter's introduction makes no secret of the fact that she finds Epictetus limited by the absence of Christian revelation.[13] On the other hand, there's no doubt that Epictetus's emphasis on the moral development of the individual supported Talbot's and Carter's preferred scholarly habits and their desire to see reading as its own end. Both were obviously drawn to Epictetus's privileging of the inner life over outward achievement. For Talbot, this justified her own feelings of distress at her daily round of social, charitable, and secretarial duties. But Carter, who once narrowly and gratefully avoided the fate of becoming a governess to the royal children, also had every reason to look with suspicion on the forms of charitable work that took up so much of Talbot's time.[14] As someone inclined to see the active duty expected of privileged Christian women as threatening to her own study, Carter welcomed the "heathen" case for the inward turn. Epictetus's emphasis on being in flow with nature, and on accepting external conditions, while indisputably in tension with ideas of humanitarian reform or equality, was a good fit with Carter's feeling of being happiest when left to her books: "The love of retirement seems to grow upon you," writes Talbot half-resentfully to her friend in 1752 (2:98). Epictetus offers assurance, which Talbot could ill afford to hear, that the activity of the critical reader might trump all the good work directed in Christian terms toward hope of a reward. More subtly, Epictetus offers Carter a justification for thinking of reading as an activity to be judged in its own terms, regardless of its moving her toward productivity or fame as a writer. It would have pleased her to find herself recognized in her later years as a shining example of intellectual virtue in herself, and as a scholar, rather than on the basis of her social or literary achievements. Like the reader who needed only to show his own tendencies, not his list of readings, Carter exhibited her progress as a reader in her person rather than in explicitly performing the lessons of the texts she'd read.[15]

Here, then, lies sanction for the way in which Carter approached the translation of Epictetus, an undertaking that remains in other ways difficult to explain.[16] If reading is to be judged in terms of mental prowess, then the amassing of erudition and the covering of literary ground matters less than the quality of training any "translatable" book might afford. The translator's Introduction that Carter wrote to *All the Works of Epictetus* makes her posture as reader legible in these terms. As an essay about Epictetus, it contains little straightforward praise for him as a figure: Carter dwells mostly on the immorality of Epictetus's endorsement of suicide and his failure to account for evil. She worries, perhaps less earnestly, over his apparent lack of interest in widespread reform, and the elitism of his speaking to only a few likeminded scholars. But she is quite open about what she doesn't understand of his text, stating that she's left certain lines untranslated even when they have previously been rendered into English. Although she makes the work of the translator visible, she equates it with finding out what is provisional in the text, developing an attitude that is critical, cautious, and well prepared for dispute.

As she makes these points, Carter exhibits the detachment and intelligence of someone who has dwelt long with this text and compared it widely to others. Ultimately, her Introduction advocates the pleasure of reading the Stoics critically. "Even now," Carter states, "their compositions may be read with great Advantage, as containing excellent Rules of Self-Government, and of social behaviour; of a noble reliance on the aid and Protection of heaven, and of a perfect Resignation and Submission to the Divine will" (xxvi). Further directions to her contemporaries make it clear that this advantage will accrue to the "attentive reader" and the "impartial reader." Her translation is not offered as a guidebook, or as a rendering of Epictetus that makes him more intelligible, but as a text that is challenging and demanding. It will require dedicated time and rereading. Talbot and Secker are in evidence here as an ideal audience. We can hear both cautioning Carter about the way Stoic philosophy clashes with Christian doctrine, and we can sense that their notes and emendations have become part of her edition. But we can also hear their approval for the life of reading that the three have shared through the making of this book, and the resonance between the life they have led as readers and the one Epictetus champions. The difficulty of the translation process props up Carter's ethical position, but also her happiness. Her many years with this text show up in her taut, careful, but pleasurable, distance to it, and the different angles from which she has seen it.

This posture of good reader depends on Carter taking pride in the work the translator does, returning to a text before updating it for her own time. Carter would have agreed, I think, with Walter Benjamin that "translation, unlike art, cannot claim permanence for its products."[17] Her interest is in bringing the Greek language into the present without securing its transparency. But Benjamin might also have agreed with Carter that reading itself is best experienced as such a process, one in which movement beyond and back to the source text takes time without becoming a simple forward movement through time. Translation opens up a temporal dimension of reading in which change can happen while remaining, in a way Carter felt deeply, a celebration of the complexities of return and repetition.

GRENVILLE'S READING JOURNALS

It's 1995 and I'm living in a shared house full of university students. We are keen readers and ridiculously confident in many ways. We make toasts to Keats, who died young. We ridicule unquestioning lovers of literature. We read Marx and Gramsci in old editions in which the library's date stamps show the trails of other readers. We ask things like: What is the theory of history? What about agency? We share books because we are poor and because it seems good that our readings overlap; this is how we find our way. I have a copy of *Mansfield Park* marked up in green pen by my friend, who once wrote an essay on landscape. She has written smart things in the margin beside all the descriptions of landscapes. I read *Portrait of the Artist* in a third-hand copy where words are underlined on every page, some in pencil and some in pen—two different readers, two different readings, two different moments. The first might have been long ago. Will we lose these as we read and reread texts on which we currently leave no marks? Or is it just that the marks we leave as digital readers are ones that we are still learning to read? This is one proposition, of course: PDFs could carry date stamps and notes, just like library books, if we were inclined to leave them. The book does not have any monopoly as an object bearing traces of its use, but it is nevertheless paper trails we are following here.

I've made the case for Carter as a rereader in time without particular reference to her handling of books. Still, it is clear that Carter's owning of paper books underpins her return to them, and the gathering of her "scraps" of translation into a book gives substance to her belief that *All the Works of Epictetus* will attract its own patterns of reuse. I turn now to an even more materially specific narrative of rereading by way of a reader committed to making his marks in notebooks he used specifically for this purpose. But I'm going to move into this territory by introducing an argument about the eighteenth-century novel, Vivasvan Soni's *Mourning Happiness: Narrative and the Politics of Modernity*, which has been in the background of this chapter since its conception.[18] Soni has much to say about happiness but almost nothing to say about the materiality of books—and this will be my point. His argument depends heavily on the trajectory of forward movement, on the idea of the book as a single-use object, which he imagines in the hands of those hunting down happiness in newly individual and apolitical ways in the eighteenth century. While there's much to admire about Soni's discussion, which plays out expertly as a question of narrative engagement, I hope to make clear how it stands to be adjusted by thinking more literally about the habits of eighteenth-century rereaders.

Soni's argument focuses on the transition between two very different understandings of narrative situated in and around the eighteenth century. The first of these understandings, he argues, is a classical one, premised on the idea of a happy life being something that can be understood only retroactively, as the sum of its different and otherwise contingently related moments. This requires the form of narrative because it is narrative that allows a community to connect up a series of accidents and discontinuities as a part of a life well lived. Such narratives, to be found, for instance, in tragedy and in funeral orations, attain wholeness once a life is over. If not, they remain accretive and open-ended, flagging the fact that happiness stays to the very end vulnerable to fate. There is no scope here for the idea of happiness felt for one moment, or viewed as an individual perspective: happiness is what pertains to a life on the whole, in general, and at the end of it all.

To this form of narrative, Soni contrasts the trial narrative, whose spell he sees as having been widely cast over modern philosophy and politics during the eighteenth-century ascendance of the novel as popular form. Trial narratives, Soni argues, involve the individual overcoming difficulty, demonstrating virtue through a rite of passage that extracts happiness from life by making it its final reward.[19] His cases include *Pamela* and *The Vicar of Wakefield*, novels in which the tribulations of characters feature almost to the exclusion of scenes representing their satisfaction. We might think of novels being all about happy endings, but in Soni's terms, trial narratives actually displace happiness from the realm of what can be written: "The trial narrative dissolves the strong relation between happiness and narrative found in the classical case, whereby narrative played a productive, enabling, and indispensible role in the judgement of happiness. One can still narrate happiness after a trial narrative, describing the reward and the process of arriving or acquiring it, but the enjoyment of the reward that is the very experience of happiness eludes narrativization."[20] For Soni, the consequences are politically devastating. The eighteenth-century novel encourages us to remove happiness from the realm of collective concern and meaningful activity. It reflects modern society's loss of the ability to think responsibly about individuals across their whole lives.[21]

Soni's argument has much in common with Small's more sociological case for the importance of the humanities to a whole life of happiness, and with her work in *The Long Life*. But while Soni suggests explicitly that we should be telling different kinds of stories, he remains silent on what seems today the much more pertinent question of where and when the reading of those stories would intersect with our other practices of well-being. *Mourning Happiness* points to the form of a narrative as a way to correct our emphasis on the individual and his struggle. But in assuming that novels, the culprits in his story, are read straight through as books, Soni bypasses more media-oriented questions of narrative consumption that might complicate the linear postponement of happiness that novels encourage as plots. What about films? What about serial narrative? Why did the narratives of antiquity Soni likes not continue their holistic function once they were available in eighteenth-century print editions? While Soni's account of the eighteenth-century trial narrative assumes the experience of the linear reader, it relies upon the caricature of the novel's main consumer as someone who moves quickly from one page and one lending-library novel to the next.

Carter's life has already offered us an example of an eighteenth-century reader for whom the linear trajectory of a book was overlaid by patterns of use and return that played into her feeling of book use being integral to happiness. To be sure, fiction was not central in Carter's library, but traces of appreciation for it are there. Carter was, for instance, a fan of Charlotte Lennox's *The Female Quixote*, and wrote to a Miss Highmore in 1752 that "there is surely some Degree of merit in a Book that will furnish one with innocent occasions to laugh, but this is by no means all the praise due to the Author of Arabella."[22] Richardson's novels also feature in her discussions with Talbot, where they appear alongside other books as texts to be reopened and debated and extracted. For instance, both Carter and Talbot read multiple editions of *Clarissa*, compare it actively to *Grandison*, and return to it in order to cite passages as advice and example.[23] "In life," writes Talbot to Carter, "one is unwilling to part with an agreeable moment because it will never come again . . . but in books one can . . . by turning back to a fine passage recall the pleasure of the first moment when one will" (1:101). Talbot was not alone in this view. These days, scholars of eighteenth-century book use are likely to agree with the case I've been making, that most readers were much more likely to extract, cut up, reuse, and read novels indexically than twentieth-century scholars of the novel's form have acknowledged.[24]

If we take such practices seriously, Soni's claim that trial narrative was modernity's "material condition" becomes shakier. For even if novels fail to narrate happiness, even if they break faith with an older model of happiness as something collectively witnessed and accretive, they may facilitate reading as a nonlinear activity independent of plot. Books differ in this respect from film, radio, theater, audiobook, and scroll, media that can be reused, but whose selective reuse is harder than that of books. As objects available and responsive to selective rereading, codex books can be extracted, abridged, and taken up in part even as they stay fundamentally whole in their original form. Taking this seriously, as I'll do to a greater degree in the next chapter, would mean acknowledging that any plot-level deferral of a character's happiness can at least be offset through a reader's use of material strategies, such as skipping ahead or selective rereading, that defy the imperatives of plot. Although eighteenth-century reading is understood by Soni and many others as driven forward in one direction by the dynamics of the novel and by the sheer availability of new text, this chapter has emphasized already that patterns of reuse emerge in this period for which neither the linear plot nor the history of religious reading provides the key. The happiness that builds up through critical reading in a life like Carter's is as accumulative as it is goal-oriented. It is not Christian. Nor is it like Pamela's. It charts, we have seen, a version of the happy life that situates books in time by seeing them as objects open to revisitation.

In defense of this position, I turn to an eighteenth-century reader of classical texts who, both more simply but also more materially than Carter, sets out to visualize on the page what reiterative reading might look like as a record of his life. William Wyndham Grenville is known mostly for his political career, first as Pitt's lieutenant in government throughout the 1780s and 1790s, and then as leader of the Whig Party in opposition in the 1800s. He became Prime Minister for a brief period in 1806 and was instrumental in this position in passing laws that finally abolished the slave trade. A privileged conservative, wedded to aristocratic values, Grenville was also a fine humanitarian politician and a serious scholar. As a prize-winning student of Latin at Oxford, he wrote a twenty-five-thousand-word essay on classical arguments around happiness. Peter Jupp, his biographer, emphasizes Grenville's standing as a scholar on these grounds and asserts: "Hardly a day passed without [Grenville's] systematic reading of the classics."[25]

During the 1790s, Grenville was at his busiest in Pitt's service. But, after marrying in 1792, he was also engaged in building up his Buckinghamshire estate, Dropmore, as a place of retreat from London and Parliament. Over the course of the 1790s, he moved all of his books and papers there and when he resigned briefly from government at the end of the century, he devoted his days at Dropmore to reading and writing.[26] I am interested in three notebooks he kept there between the winter of 1796 and the end of 1799, a period in which he was actively enjoying study without being entirely devoted to it. I opened these notebooks one afternoon in the manuscript room of the British Library where they are cataloged as "Reading Journals."[27] They are simple, attractive things, soft-bound in marbled covers and filled loosely with neat, handwritten notes on Grenville's readings of texts including Demosthenes's *Orations*, *The Life of Pericles*, and Gillies's new 1797 translation of Aristotle's *Ethics and Politics*.

Grenville's notes display his various techniques as a reader, making them valuable evidence of how a privileged eighteenth-century man handled leisure-time reading. He is fairly diligent, for instance, in his use of précis, a technique often recommended in this period as a way to internalize what one is reading.[28] We see this in his reading of Demosthenes, which he begins on October 28, 1796, with the view, he states, "of going entirely thro' his Orations" (figure 2.1). He does continue to read Demosthenes fairly steadily until February 1797. But his progress is interestingly nonlinear. On the first day, for instance, he reads one Oration. On the next, he reads another. On the next, he returns and reads the first again. On

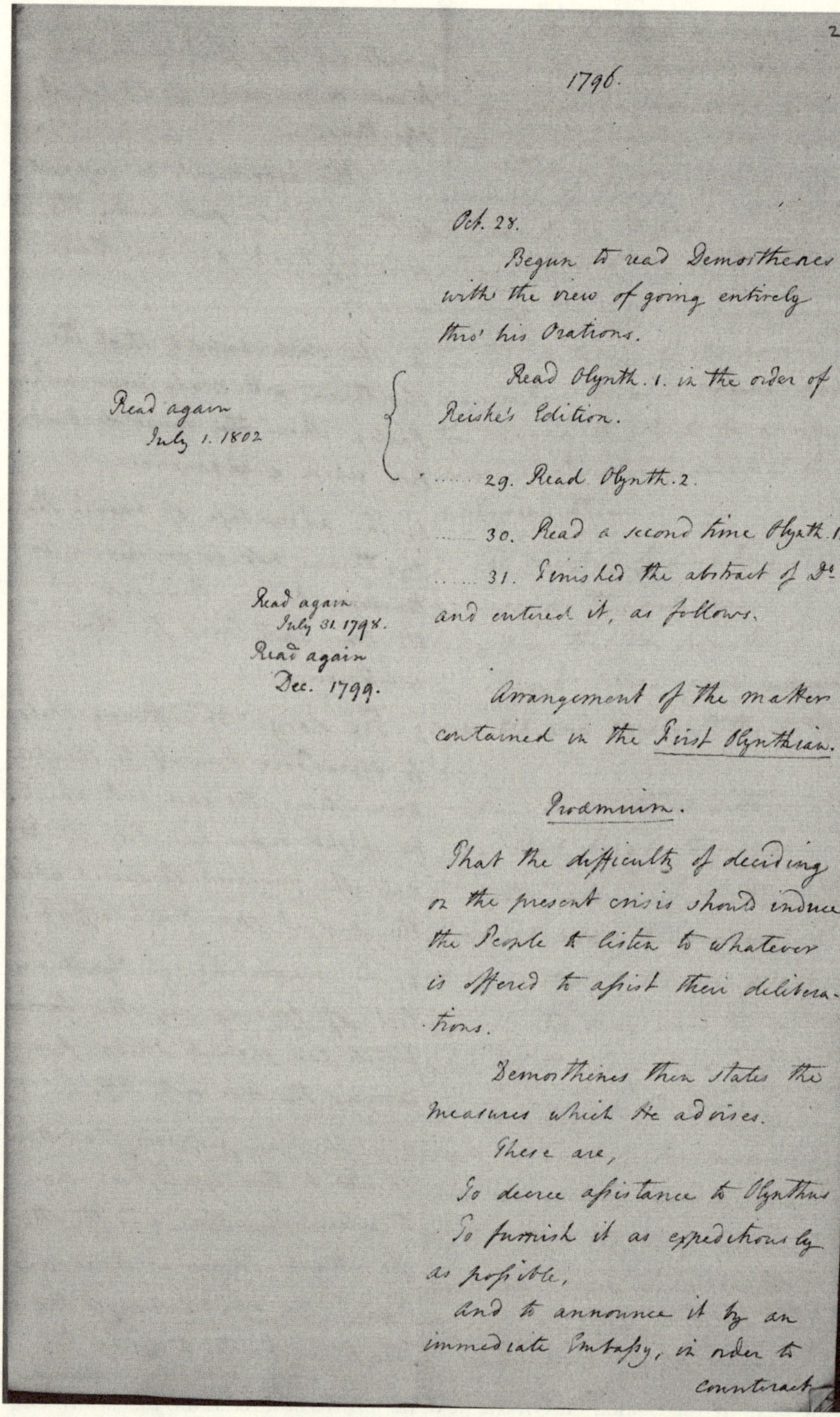

2

1796.

Oct. 28.

Begun to read Demosthenes with the view of going entirely thro' his Orations.

Read again July 1. 1802 — Read Olynth. 1. in the order of Reiske's Edition.

...... 29. Read Olynth. 2.

...... 30. Read a second time Olynth. 1.

...... 31. Finished the abstract of D^o^ and entered it, as follows.

Read again July 31. 1798.
Read again Dec. 1799.

Arrangement of the matter contained in the First Olynthian.

Proœmium.

That the difficulty of deciding on the present crisis should induce the People to listen to whatever is offered to assist their delibera-tions.

Demosthenes then states the measures which He advises.

These are,

To decree assistance to Olynthus

To furnish it as expeditiously as possible,

And to announce it by an immediate Embassy, in order to

counteract

Figure 2.1. The first page of Grenville's "Notes on Reading of Winter 1796–7 and 1798." © The British Library Board. Dropmore Papers. Add. MSS 59428, f.1.

October 31, four days after starting, he finishes and enters an abstract of the first narration.

The notebooks contain many sequences like this, all of which show Grenville reading in rounds of rotation between first readings, second readings, and his own précis of what he has read. In this context, summary involves gleaning the essence of an argument and then writing it down from memory in his own words. Like Carter, he often reads texts in their original languages, which helps justify his sustained engagement with them. But he also reads texts multiple times without summarizing them: in a later notebook from 1803, he simply lists multiple readings of the New Testament and *The Odyssey* in Greek.[29]

What is most striking about Grenville's notebooks from the 1790s, however, is the way he has folded each page, creating two columns, of which only the right-hand one is ever entirely full. He uses the left-hand column for marking new sequences of readings at a later stage. In the case of Demosthenes, for instance, the left-hand side of the page shows that he returns to the *Orations* in September 1798, going over once again what he had read the previous year. Graphically, this produces pages where reading is mapped along two axes, a vertical one that marks Grenville's zigzagging progress through a text, and a horizontal one that records, from right to left across the page, the different runs of reading that occur during the three years.The formatting of the pages suggests that Grenville was proposing from the outset a course of study for himself that would involve multiple readings. He saw reading as an activity that needed tracking in both directions, as a sequence in time, but also as repetition over time. The notebooks provide material evidence that what he had in mind for his days at Dropmore was a practice of reading as deep as it was wide.

Perhaps unsurprisingly, many of the left-hand columns in Grenville's notebooks remain blank, provocatively open to the new readings of Demosthenes forever still to come. But he does also use his left-hand columns to record several second and third readings of texts during this short period. In many of these cases, the left-hand columns work as spaces in which he updates or confirms his original impression of a text, keeping track of the ebbs and flows of his own interpretations. In February 1797, for instance, he reflects on reading the *Life of Pericles*, worrying over whether the translation might not be improved if certain words were omitted. On the left-hand column of the page, he notes, "On reading this passage again Dec 29 1790 I feel more confident that the omission of these words is the right emendation of the text. I observe that Amyot in his translation omits these words."[30] Carter would almost certainly have sympathized with the way Grenville uses questions of translation to lay out his readings over time. While committed to her Epictetus as the end result of a prolonged translation process, Carter lacked a material format other than letters in which to display her use of time. Grenville's notebooks solve this problem by creating a graphic format capturing the temporality of reading's critical dimension.

As in Carter's case, Grenville's reading performs an aspect of the happiness he is reading about. His choice of reading pertains both to a politician's mid-career quest for a different kind of lifestyle and to the way reading becomes his main instantiation of that more leisurely regime. The most concerted set of notes in his 1790s journals are a response to Aristotle's *Ethics and Politics*, which he reads in English but with great care. Grenville begins the first volume of this translation in 1799, proceeding to read and reread at the rate of twenty to thirty pages a day, and filling several columns of the notebook with his response to Book 1 of the *Ethics*.[31] What he finds there is a case for happiness, one with which he was already familiar from his days as a student, but which now confirms the inclinations that have led him to his library at Dropmore. It is in the *Ethics* that Aristotle considers happiness as the highest and only truly non-instrumental human goal, arguing for its existence in the unique form of a sustained program of activity undertaken for its own sake. It is also here that Aristotle discusses Solon's proverb, "Consider no man happy until he is dead," the phrase that features most prominently in Soni's discussion of the classical understanding of narrative.[32] It is, argues Aristotle in relation to Solon, "in consequence of virtuous exertions, continued through a sufficient length of time, a good man, completely furnished with the accommodations of life, will resume his wonted serenity; and may be pronounced happy."[33] Soni unpacks Solon's proverb slowly before settling on Aristotle's interpretation. But Grenville takes it quickly in his stride, summarizing it simply in his notes as meaning that until death "it can never be said of any man that his life has in the whole been a happy one."[34]

Grenville thus finds in Aristotle's *Ethics* an endorsement of the contemplative life of the rereader equivalent to the one Carter found in Epictetus's *Discourses*. Later chapters of the *Ethics* apply directly to Grenville's situation as they compare the role of the statesman and the intellectual, naming as superior the man whose intellect is "contemplative" and who finds "self-sufficiency, leisureliness, unweariedness" in his pursuit (X:7). These words seem written directly for Grenville, whose notes suggest that he withdrew to Dropmore in order to cultivate through a program of repeat reading the life of contemplation Aristotle advocates. "The proper good of man consists," writes Aristotle in the pages Grenville studied, "in the exercise of virtue continued through life; for one swallow makes not a summer; neither does a short day, or a short time, constitute happiness" (1:7). If performing Epictetus's advice in the domain of learning meant, for Carter, displaying one's critical and detached musculature as reader, performing the *Ethics* involved for Grenville a longer display of endurance. A notebook showing the reading of one text at many points in time could represent this kind of contemplation as a form of life.

Soni might argue, of course, that Grenville, like Carter, is no novel reader, and that he goes to Aristotle because he finds happiness described there as an ongoing possibility that he can instantiate as a reader of the classics. Richardson or Goldsmith would not have worked as authors for him to study in this way because in their novels Grenville would have found Christian happiness figured as a reward for a trial, something beyond the reaches of the reader as well as the character. But it is clear, I think, that Grenville is not interested primarily in studying Aristotle's *Ethics* as content: he had already done this as a student, and it has not produced the life of contemplation he seeks. It is the rereading of Aristotle that interests him. His performative relation to the text lies in this rereading; and it's the rereading over time that he rigs his notebooks to record. Grenville's uptake of happiness as a narrative might not have worked as well in relation to texts that did not discuss happiness, and would have been more difficult in relation to novels, but it would still have happened in terms very different from the ones Soni has in mind. For the real basis of Grenville's reading practice lies not in the content of what he read, or in a single instance of reading or response, but in the material infrastructure of the library and the books he uses as stable places of differentiated return. While Aristotle matters to this approach, it is really the technologies of the well-wrought eighteenth-century book, the notebook, and the gentleman's library that allow his commitment to making rereading jell as a good use of time, the stuff of happiness in a modern life.[35]

LIFETIMES OF READING

I have said I did not love *Wuthering Heights*. Here are some of the authors I did love early on: D. H. Lawrence, Daphne du Maurier, E. M. Forster, Margaret Drabble. I mostly stayed quiet about them because I didn't know how much I really understood them. I found samples on my mother's bookshelf and looked for more works in the little local library. The question of whether I read them right seems less interesting now that I can admit I read them mostly for the sex scenes and by skipping along to get to the narrative. Sometimes I also read them as manuals for how to live. We were in a small town, displaced in many ways by my family's move abroad. I needed books to tell me about the wider world and about life back in England, and about how to be. But books weren't very good at this, and so after a while they stopped being manuals and became the stuff of my trade, things to write and teach about. And now they are something else—maybe manuals again, when I add up the readings, because the distance from my first to my last readings really has become evidence of life having taught me something. As Small argues in Mill's case, "the mature relationship to the objects of culture that one has learned to care for and to value as a younger person carries a component of remembered pleasures that may qualify or even contradict one's response to them now."[36] The luxury of reading *Sons and Lovers* recently, after decades had passed, came partly from remembering how I'd read it the first time.

If I'd had Grenville's notebooks, or perhaps even a library in which to keep one set of copies, I could have tracked all this. If we were already more advanced in recording and deciphering the records made of our online patterns of reading, I could tell you about where I've been this last decade or so, and perhaps made a narrative of it. How many times have I searched for Benjamin's "The Task of the Translator" online? Too many to count. How does this repetition compare to my repeated return to a website that changes? I can't answer. Instead, I turn to a novel, Amelia Opie's *Adeline Mowbray* (1804), which does some of this tracking of reading over time in fictional form. Grenville almost certainly didn't read *Adeline Mowbray*, which was published when he was already busy in Parliament again, but he might have liked its larger message: read things twice, compare, and split the difference. Carter, who probably did read it, would certainly have endorsed it on the same grounds. In ending this chapter with a fiction that performs this message in the presence of the two readers I have looked at so far, I engage Soni's argument on its own terms by showing that novels themselves, cast in *Mourning Happiness* as generically unable to ever reward a reader, were able to advance in narrative form the phenomenology of happiness associated with critically rereading the paper book.

Adeline Mowbray is easily seen as a novel about reading. Its central characters are a mother and a daughter whose propensity to read badly sets the plot in motion. Editha Mowbray, a widow in her thirties, is an aspiring salon intellectual, a woman trying to make a profile for herself by cultivating a reputation for genius, neglecting household work, and consuming theories and controversies that she can try out as her own: "For her, history, biography, poetry, and discoveries in natural philosophy, had few attractions, while she pored with still unsatisfied delight over systems of morals and metaphysics, or new theories in politics and scarcely a week elapsed in which she did not receive, from her aunt's bookseller in London, various tracts on these her favourite subjects."[37] Editha radically updates in some ways a character like Sterne's Walter Shandy, whose bookishness seems such a completely male prerogative. Women, for Sterne, read quickly and lustily. They are in haste. It is men who waste time with books as fads and false systems. In giving us a woman surrounded by books, rapid in her ability to absorb and perform their ideas, Opie might already be seen as progressive.

But Editha's failure to calibrate what she reads properly to life counts against her and suggests early on that *Adeline Mowbray* is hostile to the more radical arguments in favor of female education to emerge in the second half of the eighteenth century. While Editha reads about education, her only child, Adeline, is left to fend for herself in an odd library and a chaotic house, reliant upon a grandmother who teaches her the rudiments of good housekeeping. In this sense, Opie seems to caricature the enlightened female reader, crediting the longer argument of Elizabeth Eger's *Bluestockings: Women of Reason from Enlightenment to Romanticism*, that a new generation of Romantics, including Opie, actively ignored and undercut the feminism of their grandmothers' generation.[38]

Adeline Mowbray has also been read in this spirit as a recantation of the radical theories about which Opie herself had been enthusiastic throughout the 1790s. A close associate and one-time love interest of William Godwin, and one of the friends most loyal to him after his marriage to Mary Wollstonecraft, Opie was intimate with and enthusiastic about some of the most radical ideas of the 1790s. But *Adeline Mowbray*, the creation of her thirties, can be seen as reneging on her youthful enthusiasm for texts written by Godwin, Wollstonecraft, and Thomas Holcroft. The problem here is not that Adeline loves what she reads naively: given access to "nothing but political tracts, systems of philosophy, and Scuderi's and other romances" (92), Adeline might, like Charlotte Lennox's Arabella, have developed a quixotic relation to the content. Adeline's case, however, is different. Although she is under few illusions about the way the world really is, her naiveté comes into play as her overly bold commitment to its rational reform. While Editha reads politics in the spirit of "romantic reverie," the same books, passed on to Adeline, are mined for their "rules of practice" (46).

Editha and Adeline's different approaches to books become pronounced when both find themselves bewitched by the political philosophy of Frederic Glenmurray, an advocate of the new philosophy, who writes against the institution of marriage. In this respect, Glenmurray's positions closely mirror the most radical of those expressed by Godwin in his *An Enquiry Concerning Political Justice* (1793). The argument against marriage that Godwin makes there had been publicly and easily turned against him in 1797, when he married the pregnant Mary Wollstonecraft. Soon after her death that year, Godwin published a frank memoir dwelling on her previous relationships and reiterating his objections to marriage: "We did not marry. It is difficult to recommend any thing to indiscriminate adoption, contrary to the established rules and prejudices of mankind; but certainly nothing can be so ridiculous upon the face of it, or so contrary to the genuine march of the sentiments, as to require the overflowing soul to wait upon a ceremony.[39]

Opie's character Glenmurray writes in similar terms in opposition to the institution of marriage, drawing "so delightful a picture of the superior purity, as well as happiness, of a union cemented by no ties but those of love and honour, that Adeline, wrought to the highest pitch of enthusiasm for a new order of things, entered into a solemn compact with herself to act, when she was introduced into society, according to the rules laid down by this writer" (52). In the first parts of *Adeline Mowbray*, Editha seizes upon the novelty of these ideas, going out of her way to display her own openmindedness on the issue in parlor conversations. But while Editha then pursues a disastrous second marriage in her own life, Adeline studiously and silently absorbs Glenmurray's critique of marriage, taking it to heart as a policy to be applied literally.

The folly of this position becomes clear when Adeline meets and falls in love with Glenmurray himself. Despite his willingness to abandon his principles and marry her, Adeline proposes that they enter into a partnership on the grounds of mutual consent. "Adeline, my dear child," responds her mother in horror, "little did I think that you were so romantic as to see no difference between amusing one's imagination with new theories and new systems, and acting upon them in defiance of common custom" (79). From this point on, *Adeline Mowbray* unfolds as the drama and tragedy that comes from Adeline and Glenmurray's attempt to live openly as unmarried lovers, with Adeline finding herself spurned, even after Glenmurray's death, for their scandalous relationship, and their daughter being marked as an outcast from before her birth.

It is with such morality in mind that Marilyn Butler brands Opie a conservative novelist and Gary Kelly judges her a conventional writer whose political opinions play at best a covert role in her fiction.[40] Taking up her pen as the wife of the painter John Opie, by then an established figure in society, Amelia Opie certainly pursued her nineteenth-century literary career at a distance from her youthful commitments of the 1790s. But while *Adeline Mowbray* allowed Opie to interrogate Godwin's rejection of marriage from this distance, it does not straightforwardly dismiss his politics or the influence more generally of radical writings on her own thinking. Biographical work by Shelley King shows that Opie continued to take many cues from Godwin, even after her own marriage and the decline of Godwin's popularity.[41] And recent studies of *Adeline Mowbray* suggest that the novel is best read as an attempt to balance the various positions that Opie held in her own life.[42] It is in this light, as well as because it is a novel whose cross-generational structure allows for the long life to be considered, that I approach *Adeline Mowbray* as a novel about what it means to reread one text at several points in one's lifetime.

A closer look at the text in question here, Godwin's *Political Justice*, shows that it plays directly to the question of how, and under what conditions, a reader might and should change her mind. Many of the arguments in this book are directly against what Burke's supporters at the time celebrated as the binding conditions of convention. In contrast, Godwin argues that contracts and promises foreclose the truly rational experience of the enlightened citizen.[43] His high-profile argument for the abolition of marriage is just one case among others he makes for the perversity of contractually binding arrangements. Deciding in advance upon anything, whether it is the outcome of a relationship, or the nature of a society, involves in Godwin's terms shutting oneself off to positive evolution of one's own opinion: "The institution of marriage is a system of fraud; and men who carefully mislead their judgments in the daily affair of their life, must always have a crippled judgment in every other concern. We ought to dismiss our mistake as soon as it is detected; but we are taught to cherish it. We ought to be incessant in our search after virtue and worth; but we are taught to check our inquiry, and shut our eyes upon the most attractive and admirable objects."[44] This argument, as I will discuss further in the last chapter, involves a critique of contract theory that has implications for how we understand reading as its own purpose. It suggests that a book, like a marriage, involves its reader in a relationship that should ideally be constantly tested and renewed, rather than simply asserted or taken for granted. If, as Godwin argues, "it is absurd to expect the inclinations and wishes of two human beings to coincide, through any long period of time," the point also applies to the relationship between books and readers (446).

Arguably, Opie therefore engages as well as rejects *Political Justice* as she contests the idea that any one youthful decision should be taken as binding, or even judged as correct or incorrect. Most agreements and positions reached in the novel, even by mature characters, appear as verdicts that turn out to need revision. This goes for Editha's much-regretted and ultimately overturned rejection of Adeline, and for Adeline's and Glenmurray's positions on marriage, which are quite fluid in the brief time they spend together. Both have occasion to regret and reconsider. In Glenmurray's case, his opinions have changed in the four years since his book was published: "Though I believe those which are unchanged are right in theory, I think, as the mass of society could never at once adopt them, they had better remain unacted upon, than that a few lonely individuals should expose themselves to certain distress, by making them the rules of their conduct" (179). This spirit of reconsideration also applies to romance. While Adeline and Glenmurray are depicted as deeply in love, when Glenmurray dies Adeline is paired up in a briefly happy second take with his lookalike cousin, Berrendale.

Youth is invoked routinely by Opie as a condition that leaves her characters legitimately vulnerable to changes of opinion. Of her daughter, Editha asserts early on "that the opinions of a girl of eighteen, as they are not founded on long experience, may possibly be erroneous" (80). In contrast to Jane Austen's characters, and a host of other young fictional protagonists credited with serious emotion, and sealed into their fates before coming of age, Opie's characters remain pointedly immature. *Adeline Mowbray* differs markedly in this respect from Austen's *Persuasion* (1817), where Anne and Wentworth are ultimately confirmed in their initial attraction to each other: seven years' distance only serves to prove the reliability of youthful impulses. Opie's characters, on the other hand, make mistakes from which the time of the novel gives them the chance to learn. As readers, they continue to read differently; as writers, they anticipate having their texts read anew at future points in time. In this sense, Opie's view of character resembles Godwin's take on marriage, and Soni's on happiness: there can be no verdict on processes, these writers suggest, before they are over. Character development, marriage, and happiness all involve the long summing up of discordant states and moments. Their reckoning, to which the reader's own logic of adjustment and compromise and narrative must be applied, takes time.

For this reason, it is significant that *Adeline Mowbray* covers a relatively long period by the standards of a novel. Eight years, almost the same time span Austen reckons with in *Persuasion*, pass between Adeline's meeting Glenmurray at eighteen and her dying after reconciliation with her mother at twenty-six. Both Editha and Adeline undergo radical adjustments of their tastes and positions in this time without renouncing absolutely the significance of Glenmurray's work. Adeline, to be sure, is taught the dire lesson of social rejection that comes with not having married Glenmurray, but she also discovers through the miseries of her second marriage that he was right to see flaws in the institution. Editha finds herself betrayed and financially exploited by her disastrous second marriage, and regrets the loss of Adeline. The remark by one of the novel's minor characters that "second thoughts are best," seems to speak generally for the novel's protaganists (130). The chances to revise and repeat, which present themselves at all stages of life covered by the novel, and are generally advised by its sagest character, Dr. Norberry, are fundamental to its plot.

But so too is the sense that this process of revision is one to which the reading of any text is vulnerable over time. The length of the process is underscored by the fact that Opie makes both mother and daughter reader of one controversial text. Editha's more socially palatable position as older reader does not forecast the one Adeline reaches in a more mature state. It is certainly not designed to represent Opie's own mature opinion of *Political Justice*, which, as King and others suggest, involved a strong element of nostalgia and respect for the heady days in which it had been written.[45] But Editha's presence as an older reader does help to build up a layered view of reading as something that happens through time, establishing that dimension in which a single reader can be seen to hold many different opinions about a single text. Opie, like William St Clair, seems to believe that studying the impact of any text should "span the reading of a minimum of two or three generations, as individual readers passed through the whole cycle from first reading as a child to ceasing to read in old age or at death."[46] By having two characters respond to the same text at different stages of their lives, Opie suggests that responses to a single text can be multiple and volatile when spaced out over time. In Serres's terms, this confirms that time must be one dimension of any diagram that attempts to map the movement of objects through space. The generational gap, which does not preclude Adeline and Editha from being interested in the same author, allows Opie to compile for Glenmurray's text a portfolio of competing readings that spread out in time. These readings suggest, to use Serres's analogy, why the players as well as the ball move in a soccer game. Ultimately, *Adeline Mowbray* shows that no one reader, and no one reading, will exhaust a text like Godwin's *Political Inquiry*. Glenmurray's philosophy, which has been the cause of so much strife, is never fully repudiated or redeemed. But it is shown to merit and require rereading. Only by picking this text at twenty and at forty can one begin to build up the meaning of what it says.

A generous reading of *Adeline Mowbray* would suggest that as a novel it brings that duration to life, making a narrative out of what it feels like to read and reread a text throughout one's life. Devoney Looser's *Women Writers and Old Age in Britain* has documented some of the hardships that women writers of this period had to contend with as they outlived their peers and continued to write into their old age. In general, she shows, the long life of women, including Opie, who lived into her eighties, has worked against them in terms of their literary posterity. But is this also the case of the long-lived reader? The high esteem in which Carter, the most bookish of the Bluestockings, was held in her old age suggests that readers and scholars fare better as women who can be imagined into their maturity. To suggest that reading is easier to plot as a lifelong occupation than writing might also be to suggest the aspect in which *Adeline Mowbray* is autobiographical. As

Opie's equivalent of Grenville's reading journal, it allows her to lay out side-by-side two different sets of reactions that she has to the radical texts of the 1790s: one, the reaction she has as their fan in her twenties, the other as a writer in her thirties inclined to treat books more as the stuff of her craft than prescriptions for revolution. How do these two positions add up, we might hear her asking? What is their sum? And by what logic might we understand that sum to be greater than its two parts? If there is, as her early critics observe, a desire on Opie's part to quash her earlier radicalism, this desire cannot fully account for her motivation in writing a novel that airs the radical mode of reading as a legitimate stage in a larger process of considering and revisiting books over the course of one's life.

Soni could, of course, argue that *Adeline Mowbray*, which pays lip service in its final reconciliation of Editha and Adeline to the logic of happiness, works in its own way as a trial narrative. It is only through great suffering that Adeline is given hope in the form of an afterlife, and a future handed to her daughter, whose story will not be told, but whose existence must suffice to make better worlds for women readers seem possible. And yet it remains significant, and beyond the bounds of Soni's approach, that Opie presents her own novel as something to be reread. She herself revised the novel twice, issuing it in 1810 and 1844 editions that substantially alter its original content. Anne McWhir remarks that these revisions, issued at very different points in Opie's own life, "invite us to speculate on her changing attitude to her own story, to the story she was telling, and to the telling of stories in general."[47] At the very least, Opie takes *Adeline Mowbray's* message, that one youthful reading is never enough, to heart in treating her own text as one that will support multiple sites of engagement over time.

Funnily enough, Elizabeth Carter might well have read *Adeline Mowbray*. She died in 1806, aged 89, two years after the book was published, cogent and dignified to the end. In a letter from May 1801 she writes with admiration of Opie's "The Father and the Daughter," describing it as a story both original and unobjectionable in its morals. Her enthusiasm for Thomas Holcroft's *The Family Picture*, also expressed in this letter, is even more pronounced.[48] New and radical novels continued, in other words, to hold Carter's interest and attract her active commentary. Carter's reading supports Looser's observation that much more cross-generational influence between women writers took place in the late eighteenth century than we generally imagine.[49] Opie and Carter's historical overlap also skews, however, Eger's argument for the earlier, more radical moment of *Bluestocking Feminism* as one on which later women like Opie were turning their backs by reneging on the promises of their more radical forerunners. Carter, at least, felt the connection between her own radicalism and a new generation's politics. What might she have thought of *Political Justice*? She would probably have found it hard to applaud—but she might have liked the idea of Opie, a young radical in the 1790s, returning with some cynicism in her thirties to the ideas that had captured her then. Carter's patience as a lifelong reader of certain texts and inquirer of new ones suggests that she experienced such changes as part of the pleasure—the give and take—of reading and discussing books over time. If she had any interest in Godwin, it would not have been as an author right or wrong, but as one worth rereading.

A friend told me that her father spent his last weeks rereading books he loved; more loath to say good-bye to them than the people around him. She said she didn't understand. Books don't change. You've already read them. It's the people you need to catch hold of. But if there's a different argument to be made, it might emphasize that because paper books have been things that don't change, that stick around, they have enabled us to catch hold of ourselves at earlier moments and to make a run of such moments. As the story of a life, this way of glimpsing ourselves in the mirror turns out to be strangely generous toward the process of maturity because it is reading's long duration itself that it is seen and celebrated.

CHAPTER 3

Other Times

> The book is always other, it changes and is exchanged by comparing the diversity of its parts, and thus we avoid the linear movement—the one-way direction—of reading. Moreover, the book, unfolded and refolded, scattering and being gathered back together, shows that it has no substantial reality: it is never there, endlessly to be unmade while it is made.
>
> —Maurice Blanchot, *The Book to Come*

READING IN THE FIELD

For some readers, finding time really doesn't seem to be a problem. Elizabeth Griffith (her maiden name) and Richard Griffith married in 1751 after a turbulent and loquacious courtship involving the exchange of hundreds of letters. Those letters, which they began to publish jointly six years later as *A Series of Genuine Letters between Henry and Frances*, document their different but equally busy lives—Richard's as a small Irish landowner, and Elizabeth's as the daughter of a theatrical family in Dublin. Both were informally educated and both were limited in their financial means, constrained throughout their lives by the need to earn money. During the course of their marriage, each wrote novels, Elizabeth wrote plays, and Richard published some collections of essays while he pursued business opportunities in different parts of the country. But Elizabeth and Richard also found the opportunity to read books in abundance. Their early letters are full of French and Latin quotations and references to the texts that they read and approve in tandem: Montaigne, Pliny, Seneca, Shakespeare, and Addison. In one letter, Richard writes: "I am, at present, sitting in the middle of a large Field of Barley and am taking care of the Binders and stackers: there are forty-seven Women and fourteen Men, at work round me, while I am reading Pliny, and writing to you."[1]

Richard, who never really settled down to one form of employment and whose attention remained jittery to the end, uses books very differently from Carter or Talbot. He often reads texts out of order, and generally seems familiar with texts' parts rather than their whole. He is quite up-front about this style of study being his preferred mode and openly declares himself unsuited to devoting longer stretches of time to texts. We should not think of the scene in the barley field as a compromise on a more dedicated scene of reading he'd have liked more. "In Truth," he writes of his own productions in the preface to the second edition of *A Series of Genuine Letters*: "I have never attempted any Thing which exceeded the length of a Page or two: I grow *scant of Breath*; I have not a Fund of Literature to deal by wholesale and am therefore obliged to retail my stocks and scraps; and perhaps if all Writers would confine themselves to a more laconic Method, it might save readers from misemploying a great deal of precious Time" (i.xiv). This description fits the style of his brief pieces, many of which he presents in no particular order. The reading that Richard recommends to Elizabeth, and which the prefatory material to the *Genuine Letters* recommends to their reader, entails a gentle, non-purposive browsing rather than obedience to pagination or a book's entirety. At one point Richard sends Elizabeth the seventeenth-century text, *Employment of Time*, recommending it as a collection of pieces rather than a continuous essay. "Read the preface last," he adds, confirming his taste in books as things to be browsed and sampled rather than consumed in linear terms (1:90).

Recent work in book and media history suggests that Richard Griffith was probably quite typical in dipping into and flicking through books. Brad Pasanek reminds us that desultory reading was common practice in the eighteenth century. He invokes Samuel Johnson who, when asked if he's ever done more than dip into a book, is said to have answered: "No, Sir, so you read books *through*?"[2] Pasanek stresses that the casual reading practices of eighteenth-century readers were truer than we might imagine in this perspective to the Latin word for reading (*legere*), which emphasizes "picking, choosing, selecting, collecting, and enumerating."[3] The grazing habits of a reader like Richard were self-consciously so: he describes himself possessed of "a sort of heterogenous knowledge, a kind of dictionary *literature*" that he acquired through "miscellenous [sic] learning, picked up here and there, sparga coegi, as [he] could borrow books" (xxxviii).

Pasanek's approach also confirms that books were not always treated as rigorously or as studiously as my case studies so far have suggested. The version of light reading most famously feared in the period was one in which a reader (usually a woman) gets swept up in fiction, carried away by romance. "It must be a matter of real concern to all considerate minds," wrote Samuel Pegge in 1767, "to see the youth of both sexes passing so large a part of their time in reading that deluge of familiar romances, which, in this age, our island overflows with. 'Tis not only a most unprofitable way of spending time, but extremely prejudicial to their morals, many a young person being entirely corrupted by the giddy and fantastical notions of love and gallantry, imbibed from thence."[4] In this equation, reading diverts the reader from worthier activities. The book competes for time by being more linear, and more suspenseful, than everyday life. It takes the reader hostage to the turning of one page after another. But Richard, like most desultory readers, was not particularly enamored with fiction or plot. When Elizabeth recommends *Tom Jones*, Richard reads it and writes back to say he is not impressed (1:53). Finding time for reading in his case does not rely on books being compelling in ways that make their consumption inevitable, but on their being to hand as pages to be used and accessed at his own speed. Approaching books as nonlinear and easily accessible introduces a temporal economy very different from the one that comes with thinking of them as long or compulsively forward-moving. The Griffiths find time to read, and they think of reading as opening up new temporal horizons and combinations, because the codex structure lends itself to rearrangement and sampling just as much as to sequence.

This is a material point, but it leads to more theoretical ones that become important in this chapter. For Richard's letters and essays show that his handling of books is linked to the way he treats actual events. A plot or a love affair could go one way—or it could go another. As readers, and as lovers, the Griffiths embrace their options without appearing overwhelmed by them. But they are, for all this, book readers, not internet browsers. It matters that books exist in a given form, that they are invitations, to use Davide Panagia's terms, "to think otherwise of givenness" rather to embrace absolute open-endedness.[5] For the Griffiths, and in the wider argument of this chapter, book reading brings to light a multitude of possible pathways without requiring all of them to be followed at once. Plots and arguments written and bound in one way, and yet accessible as combinations that could be different, imply that the *read* book is an entity quite different in this capacity from the unread one, which seems, if anything, to represent time as more linear than in real life. The book's fixed arrangement, its boundness and sequentiality, play into this sense of time being a field of variation and flexibility rather than lack.

In making this point I want to look quite closely at novels that feature the marriage plot as something tightly bound to conventions of the novel but ventilated by book *reading* to another kind of time. The Griffiths themselves approach their marriage creatively as readers, publishing their own love letters as things to be read out of order and opened randomly at any page. The two novels I discuss,

Fielding's *Amelia* (1751) and Frances Sheridan's *Sidney Bidulph* (1761), seem to engage with this possibility of books, and relationships, being read out of order, and back to front, or in parts, or in different combinations. They alert us to the possibilities of events being replayed, or reread, or grasped before they are fully unfolded. The time reading takes seems less linear and less restrictive in these novels when we think of them as things that can be opened and closed in different ways; and the marriages they represent seem more open, though not in an obviously promiscuous sense, once we see their structures as flexible to different futures and pasts.

LINEAR AND RANDOM ACCESS

I sit before my laptop, which is still closed on my desk, and my first cup of coffee is half-full. Over the years I've mostly moved back from writing on screen to writing on paper, so opening the laptop is optional for a while. But it is hard to resist, that first flush of information a connected device brings each morning. If I look, surprises will be revealed: emails that have arrived overnight, news stories, posts planted by my friends, calendar reminders. Even without the emails and messages, there are so many documents I could work on, starting at so many different places, a whole arsenal of eighteenth-century texts I could read, some of them completely neglected since their publication. So many times, I have asked myself after an hour online: How did I come to be reading this webpage? What was I searching for anyway? Just engaging with that machine risks ending up somewhere I never imagined. My papers, on the other hand, seem to ask for a relatively simple engagement, an engagement of one word, one line, after the next: whether written or read, they keep me on track.

This is one of my own motivations in writing and reading on paper, and it is a point that contributes to many arguments made about the characteristics of print reading. Long before it became fashionable to talk in terms of print media, Frank Kermode showed that novels could be distinguished from life on the grounds of their forward-moving plots. While we are born, *The Sense of an Ending* argues, in the middle of things, amid accidents and by chance, and guided by time only as a succession of moments, fictional characters exist differently. Their worlds have beginnings, middles, and ends; they move forward with concordance. Insisting for this reason that novels involve temporal rather than spatial organization, Kermode goes on to show that, however wide-ranging and inventive an author's representation of pasts, futures, and dream states, a novel's orientation in time follows the axis along which it is read, from left to right.[6] While Soni, as we saw in the last chapter, assumes that the reader pursues happiness hopelessly in her chase to the novel's end, Kermode argues more optimistically that novel reading moves us through time with a momentum and sense-making power that real life does not have. In Kermode's terms, the novel resists contingency. It works against the real world of "matter, material, matrix," through which reality-as-contingency remains a threat.[7]

Media historians have made the case for the reader's forward movement with similar emphasis on the reader as following a one-way path laid down by writing. For Kittler, for instance, alphabetic inscription involves linearity, and with it the follow-up movement of the reader along lines and pages. Although writing consists of components of language that could conceivably be replaced by each other, it is in his terms primarily syntactical and linear. For Sybille Krämer, it is this that distinguishes writing from technical media for Kittler:

> In the age of writing and of the book, symbolic time, by being fixed in space with linear syntactical structures, becomes repeatable and, to some extent, also moveable. What is unique about the technological era (from the gramophone to the computer) is that these technologies allow one to store "real time"—in other words, that those processes that cannot be fixed by syntactical structures and are thus not irreversible, but rather contingent, chaotic, and singular—and, at the same time, to process "real time" as a temporal event.[8]

In other words, unlike technical media, written texts encode their reality before they are accessed, immunizing themselves in this way against real accident. Of course, from a book historical point of view, mistakes get made, pages lost, and words misprinted, but the book's basic mode of operation is clear: once its units have been chosen, their order is to be followed by a reader. Like plots in Kermode's terms, pages depend in media historical terms upon reading as a lockstep formation.

It is hard, of course, to deny the basic distinction between reading a book as negotiating a course that's already set and online reading, which allows for new combinations of texts. Jerome McGann wrote, long before online reading was the norm, of the condition of "radiant textuality," allowing readers to remake books as they read them; to take them apart and reassemble them according to their own impulses.[9] In a similar spirit, Vilém Flusser celebrates the way "a new form of thinking based on digital codes directs itself against procedural 'progressive' ideologies," and Lev Manovitch declares the era of narrative over, with all digital text now laid before the reader as a database rather than a series.[10] A database offers in these terms a release altogether from spatial orientation, a memory resource into which we can dial freely from any point.[11] Even Andrew Piper, writing much more cautiously of the differences between print and online reading, emphasizes the space of the book being one in which the reader's movement involves a sense of orientation, a choreography that contrasts with the experience of internet browsing.[12] In these terms, many of the rereadings I've discussed in the last chapters could be described as assuming the stability of paper texts as things that offer readers familiar pathways along which to travel.

Yet, as we've seen with the readers I've mentioned so far, once we begin to think about the freedoms of a reader to conjugate her reading of a text in time, many practical inversions of linearity involved in book use appear. In contrast to users of the scroll, Roger Chartier and Peter Stallybrass have argued, book readers were keen from early on to navigate texts using pagination and indexing.[13] Their hands were not required at all times in the reading process. Unlike listeners of the audiobook, theater audiences, or users of VCR, book readers could single out parts of a book, return to a scene they liked, or open a book at random.[14] Scholars including Wellmon and Pasanek suggest that print reading has always taken advantage of this manipulability.[15] Recipe books, directories, dictionaries, heavily footnoted and annotated texts, Bibles, and manuals all count toward the vast majority of texts that don't encourage forward reading. Countering those who proclaim the novelty of the database, and emphasizing all these ways in which books have always been read against the grain of the linear narrative, John Durham Peters concludes: "Readers have always hopscotched around, according to their needs."[16]

Perhaps, several critics working in this book historical terrain imply, we've been far too focused on the book, and not given enough attention to the page, the unit of print upon which attention is in practice always focused. Peter Stoicheff and Andrew Taylor point out that "the book itself is never fully encountered except as an expectation or recollection or closed volume. The page, by contrast, is seen in its entirety, simultaneously. It is the constant presence, directly encountered, in the otherwise insubstantial engagement with the mirage of the book."[17] Reading by the page seems much freer, much more open to the kind of combinatorial possibility, than reading books from beginning to end. Such arguments push back at the idea that print has been a format to which readers simply conform, or that novels come with a directive to move us forwards, from left to right, front to back. But the point is not simply that print readers were (and are) freer than we might imagine to skip, or focus on, or tear out pages, to access books randomly, or to mix them up as we use them across a week or a lifetime. In material terms, it is also important that the complex machinery of the book brings linearity and other kinds of random and circular movement into contact. Matthew Brown, for instance, writes about an early American setting, where the book models "a sacred time zone that is both cyclically repetitive and linearly dramatic. An alternative time consciousness is itself nurtured by the genre of written record . . . the hand-operated codex."[18] The fact that one may enter a book anywhere, while knowing that it has an order, and that one may skip around between pages, but not truly reshuffle them, makes reading a complex activity in which an immovable order offers the reader an environment in which she nevertheless exercises choice.

A book's being bound and a story's being already written and having its end in sight are not, in this context, barriers to freedom and randomness but facilitators of more circular and open-ended journeys that its reader might undertake. This brings us into terrain that Caroline Levine covers in *Forms: Whole, Rhythm, Hierarchy, Network*, which opposes in general terms the idea that the shapes, patterns, and arrangements determine the reception of their content. There is no reason, she argues, why a plot, a classroom, or any other "form" needs to be read as closing things down or paving the way for its uniform reception. We know, for instance, of the seminar room, that spatial closure that actually opens up an hour of discussion. Similarly—Levine works lightly, by association between these realms—a novel's plot opens up life beyond and within its boundaries: "The ending's political force depends not on resolution and finality, but on repetitions that will extend past the time represented in the text. To call this closure and containment is to overlook the future implied by the text, a deliberately uncontained temporal process."[19] Levine does not invoke systems theory, which is where her argument might lead theoretically, and which is where I want to push it in a moment. But she refers, for instance, to the way in which marriage as it is represented in a novel like Gaskell's *North and South* is an opening in time as well as a closing in space: as it is read, its conclusion is received many times over, producing a whole array of reactions as the novel continues to be read at different points in history. This might also, as the cases of Carter and Opie in the last chapter suggest, describe the kind of variation that the fixity of the book and its contents supplies to a single reader who returns to it at different points in time.

Once we begin to apply such spatiotemporal models in the realm of book history, it becomes clear that thinking of linear reading and random access as materially opposing movements does not really explain the way the book orients us in time and space. The hand-operated codex as a technology demonstrates why something might be open *because* it is closed, re-ordered temporally *because* its sequence has been fixed. By this logic, partial or desultory readings and their combination depend on the substratum of the book as something that is stable in space, but has the openness to duration on its side. Combination and repetition with variation, possibilities more active in relation to a physically stable archive, characterize the freedoms that come with reading over time. It is because books are and were *there* (as Piper's *Book Was There* stresses) that they become sites of contingency as well as bulwarks against it. Although books involve horizons of spatial limitation at every level (a page is fixed, a book is limited, a library is finite), the book reader acquires powers of combination and selection distinct from those that accrue to the user of a more genuinely open archive. My books and pages may present themselves every morning as more stable than my in-box and my web pages, but in many ways this makes the way I arrange them as readings in time more complex.

LITERATURE AND CONTINGENCY

This point is relevant to the arguments I want to make now for the Griffiths' *Genuine Letters*, *Amelia*, and *Sidney Bidulph* as texts that engage creatively with the desultory and nonlinear possibilities of reading that books invite. All of these texts engage with the book as linear and foreseeable as one step in suggesting that reading practices are not. As the spatial stability of the book registers at the level of its narrative, it becomes a prompt to think about the movement of the page over time as unpredictable. At its most abstract, this invites thinking about contingency, a concept central in systems theory, as it applies to both book and narrative. In order to be helpful here, contingency needs to be distinguished quite carefully from the related concepts of chance or accident, with which it is often used almost interchangeably. In philosophical terms, contingency describes quite specifically the exclusion of impossibility and necessity from a proposition. This means that a contingent event is one that has happened, but in a way that makes it apparent that it need not have been that way: that you are reading this now is undeniable, even though chances were good that you would never do so. A tree falling in one's path is contingent in the sense that it is both unavoidable and easily imaginable as something that might have happened differently. "Something is contingent," Luhmann argues in *Social Systems*, "insofar as it . . . is just what it is (or was or will be), though it could also be otherwise. The concept describes something given (something experienced, expected, remembered, fantasized) in the light of its possibly being otherwise; it describes objects within the horizon of possible variations" (106).

For Luhmann, *Kontingenz* plays a central role in explaining what it feels like to live in the modern world. It underscores the fact that selection occurs, that we do something this way, and not that, and that we have this interaction and not another, even when the actual elements selected in combination could have been different. Parts of the law or media system or art system, as we've seen, are visible as selections and combinations. There is nothing that materially identifies them as such, no news item that is essentially media, and no book that is essentially art.

This produces a constant state of awareness that things visible as settled are also those most visible as things that could have been different.

In Luhmann's version of a post-Enlightenment world, this is how sense-making systems solicit our resignation to the world as we have made it as if it were objectively given. Contingency accounts for the coupling of our feeling that things are beyond the reach of our control or understanding with the knowledge that things are arranged by human means alone. Unlike seventeenth-century citizens, most of us today do not believe that events are preordained. We feel that global warming is not fated, that elections could have had different results, that the IMF could have different policies. But we don't necessarily have any faith that we could change these things or that their outcomes are negotiable. This feeling of living in a contingent world has been closely connected, for instance, by Richard Rorty, and by David Wellbery, to those situations evoked by literature. For literary language can also model contingency. Sentences and literary forms carry traces of the randomly accessible pool of language from which they are drawn, even as they fix on one letter and sentence after another. Wellbery has poetry in mind when he invokes Luhmann to describe contingency as the condition of literature: "Contingency is always a selection, an actualization that draws on a reservoir of other, non-actualized possibilities, a throw of the dice, an intersection. Without this selection there would be no events to concatenate in narrative series, but the selection itself—the fact that this, and not something else, happens—belongs to no chronological pattern."[20]

More generally, writes Wellbery, it is as stories take shape that we become aware of their events as random. The question of how one ended up in one's current marriage, or why one lives in the city one lives in, may well provoke a firm narrative answer, a story that tells, one step after another, what led to this outcome. But narrative used in this way de-emphasizes rather than unsubscribes from the other lives we might have led. The key here, to go back to Levine's understanding of form, is that a narrative's closure becomes the logical site of its relationship to other possibilities, both to those in its own future, and to those invoked but not realized by its own form. The more certain a narrative appears in this sense, the more it offers a sign that things might have been different. The more we explain how unlikely it was that we met this particular brilliant person to whom we are now married, the more we are likely to make visible the fact that we might never have met one another.

Wellbery's interest in contingency is language based. But in recent times we are just as likely to hear contingency invoked to describe the material conditions of disorder from which books spring, and to which they refer back. Book historical approaches have helped make *this* paper, *this* font, and *this* arrangement of words legible as a sign of the contingencies involved in creating and circulating a book and in all the physical conditions governing a text in its materiality. The choice of book itself can demonstrate the point. At an academic workshop in 2016, a group of us gathered on a far shore were asked to say which books we'd brought from the cities where we lived. The choices seemed odd. They ranged from travel guides to pornography to seventeenth-century literature (the workshop was on contingency). But there they were: we'd each plucked a book from our shelves and put it in our suitcases, and that limited selection seemed to speak simultaneously to the randomness of what we were doing and to the huge but finite collections of books and libraries we'd drawn on in making our selections. Thinking about textual materiality makes legible a trail of could-have-been otherness that flanks every single book, page, and printer's mark as it comes into focus.

My readings below lean on these applications of contingency as a concept to the literary realm, on the sense that language itself can be the scene of experiencing spatial closure as the occasion of temporal possibility, and on the ways that even closed books, and their physical components, travel through time under the sway of this logic. None of these applications of contingency to book history directly involves reading. In fact, as Leah Price shows, unread books may be subject to greater levels of material uncertainty than read ones.[21] But when reading is involved, the codex book also comes into play as a model of a literary event's contingent status. For instance, the idea that time can be spent in many ways is palpable in pages as things that must be chosen and re-chosen even after a book is in one's hands. Their flipping backward and forward, the kinds of circular and occasional return they invite, expands practically how we see the reception of any one rendering of the book's plot. Just being conscious of the book as something different from life because its pages can be read in multiple ways over time changes how we think of the fictional world being inhabited. In narratives that associate themselves with the book's material flexibility, contingency also comes into play imaginatively.

We see this, for instance, in the work of mid-eighteenth-century novelists interested in representing chance. Capturing chance in writing, Kittler has already shown us, should fail: there can be no written equivalent of the short video clip in which a baby falls unexpectedly into a puddle. Things in narrative don't happen by chance. This is also Kermode's point: novels consist of events that are plotted and fundamentally foreseeable. Laurence Sterne's single-page chapter on chances in *Tristram Shandy* (1759–67) is one of the eighteenth-century scenes explicitly designed to tease us with this insight. Most of the chapter consists of a conversation between Walter Shandy and his brother Toby on the occasion of their learning that Tristram has had his nose crushed during his birth: "What a long chapter of chances do the events of this world lay open to us!" proclaims Walter. "Take pen and ink in hand, brother Toby, and calculate it fairly." To this challenge, Toby responds that he could make no such calculation, and he swings his crutch to emphasize the point, catching Walter's leg as he does so, and then excusing himself with the exclamation, "'Twas a hundred to one."[22]

In Sterne's joke, calculation inevitably forecloses chance. Yet many eighteenth-century novels are directly concerned with the way it continues to play a role in the lives of fictional characters. Jesse Molesworth makes this reason to challenge the idea that fiction in the period was developing primarily as a form of verisimilitude. In his view, what divides the novels of Henry Fielding and Daniel Defoe from real life is that they pile up as a series of unlikely incidents, adjacent in their appeal as entertainment to the practices of gambling and lotteries so popular in the period. Reading these novels, he argues, feels more like playing a card game in which probabilities and chances emerge more as a matter of luck than like observing the confusion of real life. And yet, his reading of the scene from Sterne I've mentioned above also stresses, their plots aren't card games: they aren't really open to re-permutation.[23]

This alignment between card playing and novel reading was explored by narratives that tried to project themselves into books as unstable spaces—a fantasy realized in the twentieth century by B. S. Johnson, whose "book in a box," *The Unfortunates* (1969), can literally be read in any order one chooses. Some eighteenth-century writers experimented with such unboundedness by anticipating a material re-permutation of their published pages as part of a book's future demise. John Kidgell humorously predicts his *The Card* (1755) being picked up by children who will cut it up, stating that his decorative illustration of a card is only there so as to catch their attention.[24] Sterne introduces the blank page on which a reader is invited to sketch a picture, introducing a formal element of unpredictability to *Tristram Shandy's* sixth volume. The cocky narrator of Shebbeare's *The Marriage Act* (1754) invokes the realm of possibility to which his pages conceptually belong, encouraging readers to explore events in the narrative as if they were still open to chance: "Now in this very Place, if an Author could lay Wagers with his Readers, Thousands of Pounds might be won; but as he cannot, it may serve a Bet a White's, where the Lives of men are play'd at Chuck-Farthing."[25] Chuck Farthing—a game that involves players tossing coins towards a mark, with the player whose coin falls closest taking all the coins played and then throwing them towards a hole, keeping those that fall in—implies the narrowing down of a much larger spread of possible funds to a smaller sum. If readers were to bet in a similar way on the events to come in a novel, they would, Shebbeare suggests, have to throw a number of possible plots around in order for one to hit the mark. Asking readers to open up the narrative's future in this manner, Shebbeare strives to give his text the character of unpredictability, reimagining his sequential pages as a choose-your-own-adventure text, or a stack of papers. But even Shebbeare must admit that the author is always going to win this game; the already thereness of the book means that the text belongs to the realm of the calculable. It is only the imaginations of the writer and the reader that hold the story ajar to the realm of chance.

These fictional scenarios in which a narrative's reception is staged as being as genuinely open as a deck of cards are much less convincing, however, than the ones in which it is the contingency of reading that books model. In the years after *Tom Jones* was published authors often staged their novels as complete entities within which the characterized reader was free to move backward and forward within a closed system. The reader of *Tom Jones* is cast, for instance, as willing to turn back to the fifteenth chapter of the book to see the letter referenced in the sixteenth, or from the eighteenth chapter "to the scene at Upton in the ninth," or to compare at any one time the pages of multiple publications, such as those of the monthly reviewers, with the pages of the novel itself.[26] Minor novelists work alongside Sterne in imagining the casual disobedience of the reader who moves through the novel along her own vectors of desire and impatience. George Colman's *Polly Honeycombe: a Novel* (1760) is prefaced by a fictional letter that includes a description of the way novels are treated to different currents of reading, describing a setting in which "the third volume of Betsy Thought-less, the New Atlantis for the year 1760, and the Catalogue of the Circulating library" lie half-bound and "much thumbed and in a greasy condition" among dresses, fans, and gloves.[27] And Eliza Haywood predicts a reader all too willing to skip ahead, promising only that "the reader, if he has the patience to go through the following pages, will see into the secret springs which set this machine in motion."[28]

These scenes of fictional reading correspond to the real-life scenarios described by historians of reading, of readers opening up the pages of books providentially or by chance.[29] But they remain fictional in a way that matters for my point about the appeal of contingency. For the novelists staging the reader's movement in these scenes anticipate the book's construction as a complete printed and bound object as a precondition of the wayward tendencies of the reader who accesses its pages, however randomly. This can be seen in *Tristram Shandy's* famous scenes of digressive reading, which involve Tristram sending his fictional reader, "Madam," back to an earlier chapter that she is assumed to have read poorly the first time.[30]

It's easy to forget that in order for this possibility of "Madam's" nonlinear reading to arise at all, *Tristram Shandy* must first exist as a book with a certain number of contiguously bound pages. In other words, the novel must be imaginatively formed in Levine's terms. It's not until it has been extracted from all the papers—the sermons, the unpublished manuscripts, the legal contracts, the scrolls, the medicine wrappers, and the notes—that circulate, both in Tristram's and in Sterne's worlds, that it can be referred to as a book to which one has random access.

The same representation of a book is involved in Tristram's claims that he will digress from the path of his narrative in the pages to come. Here Tristram announces that he has just surveyed what he has written: "Upon looking back from the end of the last chapter and surveying the texture of what has been wrote, it is necessary, that upon this page and the five following, a good quantity of heterogenous matter be inserted, to keep up that just balance betwixt wisdom and folly, without which a book would not hold together a single year" (559).

Even as he deliberates upon the ballast his novel needs in terms of content, Tristram assumes pages that seem already to have the form of a notebook. At another point he glosses over the difference between manuscript and print pages by suggesting that he shares with his reader one paginated environment and the same freedom to cut multiple paths through it while leaving it unchanged: "I have dropped the curtain over this scene for a minute,—to remind you of one thing, and to inform you of another. What I have to inform you, comes, I own, a little out of its due course; for it should have been told a hundred and fifty pages ago, but that I foresaw then 'twould come in pat hereafter and be of more advantage here than elsewhere" (128–9). Tristram refers to the spatial layout of the page—divided now, by a chapter break—as well as to the arrangement of pages he has previously written in relation to each other. The conceit places Tristram and his reader inside the same book, where both appear at liberty to manipulate its paginated form. Yet these references to the extant book remain fictional. They were written on loose sheets of paper before Sterne's first volume was published.

This is not to say that Tristram himself should not be imagined as a purveyor of books, a manipulator of the codex machine. As Karen Harvey has argued, plenty of eighteenth-century male writing was being done in the bound diaries, logs, and registries that presented themselves as blanks to be filled in, pored over, and even edited, just as they were also seen as finite and fixed permutations of pages.[31] And even when documents were printed, they were often printed as frameworks (journals, diaries, etc.) within which writing was then produced: the possibility of layout and form being there first and creative writing second is not illogical.[32] Locating Tristram as a writer in this environment helps to suggest just how important turning backward and forward in an already paginated environment was to his own sense of contending with his family history. But even so, lining up *Tristram Shandy* as an already paginated text with the unbound pages on which Sterne himself wrote can only be done imaginatively. There is no literal connection between the materiality of the single volume Sterne's first readers would have held and the pages on which references to this volume were first written by his hand. From the point of view of the aspiring writer, Sterne, like many other self-reflexive authors, contributes to the fallacy of the reader being at liberty to turn pages only by acting disingenuously on the premise of his manuscript being already printed and bound. He does so, I'm suggesting, because this artifice allows him to bring contingency into view. The environment of the book that is settled—this way, and not that—is the perfect setting to open up to the reader a portfolio of other endings that might have done just as well.

Jane Austen's *Northanger Abbey* might be used to make the same point. Her narrator also invokes the pages of the book in the reader's hand. She does this in the last chapter of the novel, as Catherine's marriage to Henry seems to hang in the balance. Austen's narrator observes here that "the anxiety" she has cultivated about whether her protagonists will marry "can hardly extend, I fear, to the bosom of my readers, who will see the tell-tale compression of the pages before them, that we are all hastening together to perfect felicity."[33] Pointing to what is now the foreseeable moment when the book will be closed and set aside, and to the inevitability of its happy ending, this narrative commentary forms part of Austen's effort to train smart readers to connect the book and the plot as limited frameworks, set off in their formal trajectories and closure from real life. While Catherine has been chastised throughout the novel for taking gothic fiction too seriously, Austen's readers are being commended here for knowing the world of realist fiction and its limits. The book mediates between the world of real chance and the novel that moves inevitably toward its conclusion.

But like Sterne, Austen is embracing the idea of the book proleptically because it allows her to build up a model of contingency. It is here, where we are most securely tied to the mast of the marriage plot, and bound by the fixedness of the book, that she invites us to imagine other kinds of lives and other kinds of readings. Real openness (say, in the form of a choose-your-own-adventure novel or a dating app in which many possible dates were displayed) would not work nearly so well. It's when the marriages in *Emma* and *Mansfield Park* are settled that Austen exposes her reader to his or her imaginative freedom. What did Emma say in response to her lover's declaration of feeling? "Just what she ought, of course. A lady always does."[34] And when do Fanny and Edmund finally marry? Austen declares wryly she must "purposefully abstain from dates on this occasion, that every one may be at liberty to fix their own."[35] Austen's trademark may be the creation of plots that seem to have nowhere to go but toward a particular romantic union, but her real skill is to make this artificial closure of the plot a condition of our raised awareness that other combinations of events were possible. In this respect she depends on the book's materiality as closed *and* open as one that serves her narrative project well.

Austen also anticipates Levine's point about the way in which *North and South* can be read as open to the real vagaries of reception precisely because it is a closed fiction. For both Austen and Levine, the single plot form activates the multiple interpretive possibilities awaiting the book. If no dice can be tossed to determine *Mansfield Park's* possible endings, then at least the caprice of the reader can be claimed as the patternless field into which the closed novel opens out. The relation between hermetically settled marriages and the uncertainty of their reception is one Paul Fleming identifies as the relation of the anecdote, with its absolute brevity, to the multiple interpretations of the characters who encounter it. This construction, he argues, involves making "contingency and necessity share the same space."[36] Austen connects the economy of marriage plots with the reader, who will treat them most lightly, and treat them as most contingent, because they are so economically settled. It is in this spirit that Austen exposes her plots to time as books. The page becomes an interface between the domain of closure that the plot and the closed book share, and the world of accident and uncertainty into which plots and books emerge once they are read over time. Turning the page, one is hastening toward the inevitable—but one is also engaging with the possibility that no sequence is absolutely necessary, and that the sense that things might have been otherwise will grow, not vanish, once the end is reached.

AMELIA'S BEGINNING WITH THE END

If Austen sees a connection between the bound book and its movable parts on the one hand, and the artificial closure of the marriage plot and its relationship to uncertainty on the other, she is not the first to do so. Nor is it surprising that novels that end in marriage should invoke in some way the more complex lives of their eighteenth-century readers. In reality, marriage was rarely the end of the story in the period; its fictional appearance as such was every bit as evidently artificial as it seems now. For most eighteenth-century women, Ruth Perry argues, consanguineal structures were much more likely than conjugal ones to ensure security and intimacy.[37] In historical terms, eighteenth-century marriages were insecure, often unhappy, and frequently short. Marriage rarely lasted a lifetime. Early death made remarriage common, and desertion and separation were likely. A quick survey of the readers I've mentioned so far proves the point: Talbot and Carter and Austen remained unmarried, Opie spent most of her life as a widow, and Fielding, Lackington, Turner, and Godwin were all married twice. Only in fiction, and as a trademark of fiction, was marriage associated with a definite kind of resolution.[38] Here, from the 1750s onward, argues Lisa O'Connell, marriage spelled out with increasing force the end point in a novel's trajectory, signifying the coalescence of narratological and interpretative trajectories, and confirming "simultaneously social status, states of feeling, Christian virtue, and moral worth."[39] In other words, it was at a point in history when marriage was experienced as provisional and unhappy, satirized widely, and postponed and repeated by many readers in their own lives, that it emerged as the trademark of realism's linear treatment of plot.

I make this point as backdrop to the readings that follow, readings of two novels that connect fairly boldly their representations of marriage to the codex book as a model of contingency. Both novels experiment in the years before Austen was writing with an idea that she then works up: that a narrative firmly concluded could be a format for dwelling more ethnographically or realistically on different plots. Like Austen, both Fielding and Sheridan sense that the physicality of the codex book colludes in this awareness. Fielding's *Amelia*, for instance, begins unusually with a couple that is already happily married and whose fate, we are told up front, is to remain so. In his review of Fielding's last novel, John Cleland points out how rare this move is: "The author takes up his heroine at the very point at which all his predecessors have dropped their capital personages."[40] While Fielding sometimes refers metafictionally, for instance in *Tom Jones*, to the pages of chapters of his novel, here it is the settled marriage between his chief characters Amelia and Billy that allows him to flag the already-thereness of the novel: we are told abruptly that we are in a novel and that no real surprise or chance awaits us. Fielding also plays here with the common belief that married life would offer no narrative suspense as a topic. *The Female Quixote* (1751), a novel preoccupied with the failure of modern marriages to make good stories, underscores this feeling when Lennox's heroine, Arabella, makes the mistake of asking after the life story of a married Countess: "'When I tell you,' replies the countess, 'that I reciev'd the Addresses of my Lord—through the recommendation of my Parents, and marry'd him with their consents and my own Inclination, and that since we have liv'd in great Harmony together, I have told you all the material Passages of my Life, which upon Enquiry you will find differ very little from those of other Women of the same Rank, who have a moderate Share of Sense, Prudence, and Virtue.'"[41] *Amelia*, however, seems determined to answer Arabella's question differently, and Fielding seems determined to suggest that there might be more to be said about happy marriage than the Countess admits. Although he has laid his most important cards on the table by bringing the plot of his novel into view at one stroke, Fielding now insists that uncertainty can still be his topic. He sets out to write, he claims, about "the various Accidents which befell a very worthy Couple, after their uniting in the state of Matrimony."[42] Given the fact that we know how this novel is going to end, what Fielding's narrator is talking about can already be described as contingency. Things are one way—neither the worthiness of Amelia and her devoted husband Billy Booth, nor their partnership, will come into question in the course of this novel—but his whole story will flag how they might have been otherwise.

On this ground, *Amelia's* romantic plot, as foreseeable and immune from chance as the book itself, competes with all the alternative forces of desire, possibility, and chaos that Fielding opens up to view through the window of this overtly settled marriage. Marriage, on the one hand, appears as routine. Billy at one point describes his life together with Amelia as "one continued series of love, health, and tranquillity. Our lives resembled a calm sea" (141). The narrator backs him up with descriptions of married life that are often deliberately non-eventful. Amelia, for instance, is at her best "while she was dressing her husband's supper with her little children playing around her" (496). Scenes of Billy and Amelia playing with their children on the rug, having discussions in bed at night, and participating in routine mealtimes make for some of Fielding's most original portrayals of middle-class daily life.

At the same time, randomness and disorder interest Fielding more here than in any of his other novels. Drawing on his experience as a magistrate, Fielding famously uses *Amelia* to display urban chaos, the injustice of the legal system, emotional betrayal, and the unfathomable corruption of London as a city. Financial ruin, imprisonment, and infidelity threaten even the morally upright. Several critics have discussed this juxtaposition of harmony and disarray, determination and randomness, in *Amelia* resulting in Fielding's failure to deliver his reader a properly successful novel. The world of London and its legal system as Fielding represents it is too much a world of chance to be legible—of Humean chaos, as Molesworth describes it—while the fate of the marriage he portrays is too highly determined to make a good romance.[43] John Zomchick describes Fielding suggesting that "corruption in the public sphere and honest affection in the private are dialectically necessary representational antithesis," while Terry Castle reads *Amelia* as upending more contrarily its own normalizing ideals by "insinuating in the place of moral certainty, a tropology of ambiguity and complexity."[44]

Yet Fielding's accomplishment in writing *Amelia* is to show how randomness and overdetermination can be conditional upon each other. In looking at the compound of liberty and necessity in the novel, Adam Potkay suggests that it is because of the control passions seem to have over characters that they are shown to be at liberty to direct and work with them.[45] A complex version of this connection is made is when Billy, relating the facts of his happy marriage, returns to the moral fray from which his union with Amelia has protected him. This encounter structures the first three books of the novel, during which Billy and Miss Matthews, Billy's ex-lover and now a suspected murderer, find themselves in Newgate Jail and contrive to have themselves locked up together so they can exchange life stories. In this literally closed context, a jail cell in which the story appears to have no room to go anywhere, Billy describes his courtship of and marriage to Amelia and the birth of their children: "During my first year's continuance in this new scene of life, nothing, I think, remarkable happened; the history of one day would, indeed, be the history of the whole year." When Miss Matthews insists that it really is the history of that day she wants to hear, Billy relents in these terms: "If you command me, madam," answered Booth, "you must yourself be accountable for the dullness of the narrative. Nay, I believe, you have imposed a very difficult task on me; for the greatest happiness is incapable of description" (146).

This scene quickly turns into one of adultery, with Billy and Miss Matthews spending a week in bed together behind bars. But Fielding's representation of adultery constitutes neither an erosion of Billy's love for Amelia, nor a correction to the facts of the romance as he has related them. The settled marriage activates the uncertainty and possibility of the affair without compromise to its own validity. Put more abstractly, Fielding presents a non-dialectical juncture where narrative closure and the wider field of contingency coexist. This is not a causal relationship: it is not because of Billy's description of his happy marriage that he has opened himself up to adultery. Rather, in this sequence of events, where it is the relation of unqualified happiness that makes chance evident, Fielding approaches the scenario Wellbery describes, in which contingency emerges when we are asked to narrate how we ended up in our current partnerships. It is only by way of contact with the realm of chance that we can explain what has been, and remains, fixed. In the case of *Amelia*, the security of the main characters' happy marriage allows for the description of the realm of contingency to which this marriage, undiminished as a given, still belongs—and which its closure as a narrative affirms.

It matters here, as the present tense of daily life comes into play again in the novel, that Billy and Miss Matthews tell their backstories and conduct their affair in hours already deducted from their everyday lives. Their time in jail is effectively dead time. The stories they tell there have already run their course. They cannot be changed. But they can be retold in whatever form the protagonists see fit. All of this contributes to the sense that their brief portion of life together—as the life that might have been but is not—is allied to reading as an activity that adds a conjectural dimension to experience rather than being docked directly from the quota of hours spent, for example, with one's legitimate family. Fielding approaches adultery more like late-night rereading than a restructuring of past life. Its temporality is less in conflict with daily routine than we might imagine. But it does bring us into contact with that world of chaos from which Miss Matthews arises and returns as a condition of the marriage plot. It becomes the very structure of *Amelia's* domestic narrative, its givenness and its routine, that allows this sideways peek at other less sustainable and secure scenarios.

The argument I'm making here, for a novel about a happy marriage as a foregone conclusion as one that is especially well qualified to represent itself as porous to reordering, can be contrasted to the argument made by Mark McGurl for the connection between twentieth-century fictions of marriage and the consumption of books. McGurl reads the marriages represented in *Mrs. Dalloway* and *The Portrait of a Lady* as choices into which Clarissa Dalloway and Isabel Archer are uncomfortably locked, with the other partnerships they have had surfacing as tragically lost opportunities. The feeling the novels produce, he argues, is the one a consumer tries to avoid, for instance by exploring each purchase as a form of open choice. Because Isabel Archer's and Clarissa Dalloway's marriages take the form of opportunity cost, they stage the kind of loss that haunts us when we must spend money one way and not another, or read one book and not another. Just as we wish Clarissa Dalloway might have married both her lovers, we also wish for ever new horizons of instant gratification. Amazon becomes in McGurl's account the late-twentieth-century fulfillment of this fantasy, allied to contemporary fictional forms that release readers from marriage as monogamy. Amazon restores to the novel its latent form as a choose-your-own-adventure book.[46] But in the eighteenth century, I'm suggesting, novels inhabiting the book modeled a different way past this impasse by suggesting in medial terms that an environment allied with fixity might also be one well equipped to supply choice. As readings, multiple fictional scenarios do not crowd out alternatives in the way lived ones do.

SIDNEY BIDULPH AND THE TWICE-TOLD MARRIAGE

Let me pursue this case by moving to Frances Sheridan's *Memoirs of Miss Sidney Bidulph*, a novel that can be described as a series of "what ifs." *Sidney Bidulph* is presented as a journal of a young woman who writes to a friend of a marriage that might have been but wasn't, another that might have worked but didn't, and another contracted on the basis of a death which might have occurred but didn't. Sheridan's commitment to conjectural thinking is evident at sentence level, with Sidney regularly reflecting on what might have happened had she married Faulkland, her first suitor, rather than Arnold, who becomes her husband. Sidney's lack of commitment to one set of events is not just a symptom of the right man having been substituted for the wrong one (in fact, it's never clear that this is the case), nor does it suggest her lack of commitment to the course of events in which she is actually swept up. Even at the early point where her marriage to Faulkland seems happily set to go ahead, Sidney talks about it as an "if" when those around her take it for granted: "If you should be married! Said my brother; I know of no possible IFS, unless they are of your own making."[47]

The slip between cup and lip happens when Faulkland is exposed for having had a misguided affair in the weeks before meeting Sidney. Learning of this, Sidney and her widowed mother decide the marriage would be unseemly and call it off, urging Faulkland to marry his pregnant lover instead. Arnold, the man Sidney is then paired with, proves a fine alternative—until he is caught after two years of marriage having an affair. By this time, Sidney appears the victim of men determined to lead multiple lives, men distracted from their own plots by eligible women with whom Sidney herself becomes through her own logic of "what if" too easily interchangeable. Faulkland, reappearing at the point where Sidney and Arnold's marriage is disintegrating, makes this kind of swapping around explicit when he pretends to elope with Arnold's lover, Mrs. Gerrarde. Yet, despite its thematic presence, promiscuity is not where contingency plays out in the novel. Adultery proliferates here as one key to the atmosphere of substitution that Sheridan cultivates, but one less important than the version of multiple fidelity Sidney commands by insisting her life be read as a book.

There is a foreshadowing of Sidney's and Arnold's different approaches to marriage in the scene where Arnold proposes. Here Sidney's existence is associated closely with contingent acts of reading while Arnold's philandering is signposted as a much more compulsive and teleological form of movement. Of Arnold's proposal, Sidney writes: "I was sitting in the little drawing-room, reading, when he came in. . . . The book happened to be Horace; upon his entering the room I laid it by; he asked me politely enough, what were my studies. When I named the author, he took the book up, and opening the leaves, started, and looked me full in the face" (80). The book here "happens" into place and is lightly set aside. Arnold opens it pages, seemingly at random. As a center of gravity, it also ventilates the room to things being otherwise. But Arnold seems uncomfortable with it in exactly these terms. His attention turns to Sidney's embroidery, a rose in the center of a little fire screen she is making, and asks if she wouldn't rather finish this than read Horace. Sidney, nettled at the question, declines to answer. Although Arnold's suggestion is clearly an affront to the woman reader, in this case the comment rebounds on him, the man who will find female sexuality hard to ignore. For her, the screen is practical and necessary, something she will return to later in an hour of financial need as an illustration of her practical skill, while the book is pleasurable and empowering as evidence of her own power of choice. For Arnold, the embroidery is an emblem, a "rose that blushes with neglect in its frame." Like the embroidery, his story revolves around a vortex bounded by a frame; his romantic trajectories also have nowhere to go. His story is opposed by the emblem of the open book, toward which Sidney's preference remains unshaken, and whose open pages seem poised as a question in Arnold's hands. Will Sidney wind the story back toward her feelings for Faulkland or move onward into the uncharted territory of marriage? The book marks the bouquet of possibility the story presents us with, while the unfinished needlework points toward what operates in this novel as the simple gravitational pull of the male protagonist toward the site of his own desire.

The book appears again in this role as moral emblem of multiple choice in Sidney's life. Two years into her marriage to Arnold, Sidney is asked by a friend "whether Mr. Arnold was not once near losing the happiness he now enjoys" (125). The curious Lady V— refers to rumors of Sidney's earlier engagement to Faulkland, a character who is about to reappear on the scene. But Lady V— does not force Sidney's confession: "If my curiosity is improper, or if there was any particular motive to this disappointment of my kinsman's hopes, which you don't choose to reveal, forgive my inquiry and think no more of it; but take up that book and read to me while I work" (126). The book features here as form, not content. As a screen granted to Sidney to hide behind, its shape implies that she might dial into her own story as she wishes, move backward and forward in its sequence, and pick up and put down the story as she chooses to tell it. The book is not the modest option, but the bold one.

These readings might seem a little disingenuous were it not for the more general sense in which Sidney's own narration of polyamory is associated throughout the novel with the shape of the book. Unlike the better-known epistolary heroines of the period, Sidney shows no particular commitment to letter writing. Despite her hefty output, she does not procure ink, secrete papers, or discuss the logistics of sending her reports to her friend, Cecilia. A day at home sees her reading or engaged in needlework, never visibly writing the words we are reading (140). Sidney's writing also makes very little discernable difference to the actions she endures. She seems rarely to expect answers to her letters or worry about their going astray. The doggedness with which she continues writing what she describes as her "journal," a doggedness worthy of Clarissa or Evelina, is connected to the value of the coherent document within which the episodes of her life will be read by a future generation of readers. Much more than Austen or Sterne, Sheridan makes the conceit of writing in a book and as a book convincing: the loose sheets of Sidney's papers seem always to have been bound, and yet still to be unfolding.

This effect works partly because Sidney's approach to journal keeping also reveals much about her attitude to life as something to which she submits without any overt attempt to control events. She greets Arnold's abandonment and return as well as the surprising reversal of fortune that occurs when an unknown relative rescues her from poverty with patient resignation. Without being rebellious in any obvious way, the "placable" Sidney is resilient, though, in her consciousness of alternative realities. Her response to her surprise inheritance from her cousin, for instance, is to accept it with gratitude while questioning whether it is real, emphasizing in her report how easily the money might have gone to her brother: "'Tis all enchantment! I am afraid my old kinsman is a wizard . . . I have been talking to, and examining my servants, to see if they are real living people, or only phantoms; I look at, and handle the rich furniture of my apartments to try if it be substantial!" (370). To think otherwise of givenness is Sidney's habit. Even when she proudly convinces Faulkland to marry the long abandoned mother of his child, she acknowledges quite openly that if he could still have married her "a very few years would perhaps have disposed me to return Mr Faulkland's . . . passion" (335).

To suggest that Sidney fails to act on her real desires would be to psychoanalyze her more deeply than the novel allows. More constructive is to see her as personifying contingency as Luhmann attributes it to modernity at large. For Luhmann, as we saw, contingency accounts for the coupling of a world that feels beyond the reach of control or understanding with our knowledge that things could be arranged differently—that they are not ordained, fated, or controlled from on high. In *Social Systems* Luhmann describes modern life as "a wholesale concession that it could always have been otherwise compensates for the baselessness on which structure is acquired" (111–2). We don't believe, to go back to Wellbery, that we were fated or obliged to end up with our current partner, but, at least from within the relationship, we generally want to take it as a given. Sidney's love life, which sees her paired morbidly and mistakenly with Faulkland at the novel's penultimate turn, heightens this kind of awareness because even as she is exposed to a range of jarringly juxtaposed realities, she writes in painful consciousness of their contingent status.

The material form of the journal, treated from the beginning as a bound book rather than a pile of letters, helps keep alive this sense of the different routes Sidney's life might have taken. Even when she reports herself most deeply content with Arnold, Sidney recalls her relationship with Faukland as safely out of reach but integral in its earlier form to the architecture of her narrative. The structure of the codex book allows earlier and later points in her story to exist at a decorous distance from each other, but in a proximity that is quite different from the one mere historical sequence would suggest. Offered up as a series of readings, the book allows Arnold and Faulkland to share Sidney without ever directly competing over her. The novel's multiplicity of true relationships, combined with the fact that none of them dominates the directionality of the plot, makes for a version of monogamy ventilated by the paths of comparison and return allowed by the motility of the bound page. The novel becomes unusual, not so much for the originality of this suggestion, but for making it explicit. Sidney's loves are separated by the time that steady reading of her journal would require while being offered up as proximate realities the reader is in fact invited to access differently.

Sidney and Faulkland's eventual marriage, upon which the latter insists at the point when he is half-mad with the belief that he's killed his wife (he hasn't), is itself both a moment of comparison between a beginning and an end, and a path of return. Rather than being a resolution, it is a gothic farce in which Sidney participates with her usual equanimity but no good feeling. Better, we sense, would have been for her to return to the opening pages of the story and begin again with the betrothal, than to follow the course of action that has led us back to it in a sullied state. The idea of progress mocks the way reading goes nowhere—Kittler reminds us—that an author hasn't already been. Sidney and Faulkland have no such luck in being allowed to begin again, but readers do: if we are to experience their union, we need only return to the point in the book where it was first promised.

The novel then, like the story of Booth and Amelia's marriage, is already there, unfolded in time. But the sequence of events in a novel is not nearly as programmed as the numbering of its pages. In Sidney's deviance from the marriage plot we see why it ties the book form to contingency rather than simply to choice. While sequence exists as the momentum to which Sidney submits, and which she reproduces as a journalist of her own life course, Sheridan cultivates a readiness to go back, to begin again, and to think conjecturally of different options, a readiness that keeps her book open. No deck of cards allowing genuine reshuffling, but also no game of chance in which the results are rigged toward a happy end, Sidney's life is simultaneously a lesson in bad luck and an invitation to feel how easily things could be different. If contingency shows us something about the spirit of modernity, here it also shows us something about modernity's close relationship to the book.

THE GRIFFITHS' MARRIAGE BY THE BOOK

I am leafing through pages again, preparing for teaching. Only a few minutes left before class begins but time to turn back and forth a million times, to speed the novel up, and go back to the quote I need before the clock chimes. I have been in this situation too often to panic, and I've concluded that I actually spool through familiar books in these moments as a way of controlling time rather than succumbing to it. It's also then that I feel how many points and connections might be made, anticipating all the turns the conversation in class might take, turning down the corners of pages we might or might not come to. The book is all there, and the way through it is straight enough, but I will never again take that path. For me, now, every time is new and every flip of the page a choice on a minor scale. This feeling could not be reproduced in electronic form, in a text that wasn't whole in the first place. But it could, I want to show in the last section of this chapter, be fed back from the book into an understanding of a real set of events.

Elizabeth and Richard Griffith were well known as a couple in the 1760s and 70s. Once *Genuine Letters* began to appear in volumes and editions six years into their marriage, readers were given access to selections of their correspondence. Their letters were read from the outset as real and not as an epistolary novel. Interest in them was high: the first volumes went through multiple editions even before the last volume appeared in 1770. That popularity is also fairly easily explained: Richard and Elizabeth are lively, witty correspondents, caught up in genuine struggles over power and their future, and unusually open about the various forms their intimacy could take. Elizabeth, in particular, writes with energy and strength about her own position as a woman, and about the importance of women's education. "I interdict you," she writes in an early letter to Richard, "from the unjustness of any satire against our sex, till you have, by a proper and more liberal education, given our noble and ingenious natures fair play to exert themselves. Do this, if ye dare, ye imperious tyrants, and ye shall see, how small we will make you" (1:48). Such exchanges display Elizabeth as highly literate but not privileged, unconventional but careful about her relationship with Richard. During the early years of their correspondence the question of marriage could hardly be raised, so impossible did it seem that they could fund their lives together. While Richard lobbies for an erotic relationship—"My desire points North, in the direction of your Chastity" (1.13)—Elizabeth writes boldly of other kinds of love: "That I love you, I own, and confess it more freely, since I find I have, thank God, sufficient strength to acknowledge it with safety; for I am glad to find, I do not love you better than myself, and, tho' I would cheerfully sacrifice all that is perishable to me, for you happiness, I shall take care to preserve that part of me, which may make you, at some time in your life, not ashamed of having loved me" (1.46). In response, Richard opens up to her about his friendships with other women, his preference for the company of women over men, and previous relationships he has had. At one point, he recycles a poem he's written for another lover, and sends Elizabeth jewelry he'd given to and had returned by a woman he'd once loved. But he's charming too, particularly in his support of Elizabeth's gifts as a writer.

As we saw earlier, Richard is fairly candid about his own short span of attention, and he presents himself in *A Series of Genuine Letters* as unable to follow the thread of an argument or a course of learning along conventional lines: "My writing, like my life, has been ex tempore, and with as little parsimony. I have sometimes crouded as many hints into one letter as would have served a French wire-drawer to frame a dozen essays out of. I have lived with precipitation, and all my oeconomy has been for the future: I have many subjects in contemplation, but never proceeded further than a few minutes; I have not patience or servility to trail a thought in leading strings" (1. xliii). Richard's unapologetic enthusiasm for diversion and tendency toward "precipitation" can be explained in part by his reverence for Sterne, the one novelist for whom he expresses real enthusiasm. While Thomas Turner, who wished to spend longer with books, could make nothing of *Tristram Shandy*, Richard, satisfied to come at them when he can, models his instincts as a reader and writer on what he takes to be Sterne's commitment to disconnection. In 1770, he published a fake memoir of Sterne, *The Posthumous Work of a Late Celebrated Genius*, in which "Sterne" reflects on the writing of *Tristram Shandy* as a complete work of chance, lacking any thread or plan: "After this careless manner," writes Griffith channeling Sterne's voice, "did I ramble through pages, in mere idleness and sport."[48]

In Richard's individual letters to Elizabeth, he skips playfully between topics, inserts quotes in a fairly slapdash way, and frequently closes his address in mock haste. His two essay collections make this absence of connection between topics of discussion their main theme. The second volume of his essays, *Something New* (1770), is introduced, for instance, with a description of two friends, one who is writing a book and one who has dropped by and "met with these papers lying on the table."[49] The friends get into an argument, with the visitor denying the principle upon which Griffith has claimed to be working—"that thought was free, and that the mind can think without a chain, and can summon or dismiss ideas, as arbitrarily or capriciously as it pleases" (2:3). As a result of this argument, the friend undertakes to read the work at hand in order to prove the chain of reflection running through it and to prove, as Richard scoffs, that the human mind "strikes two because it has sounded one before" (2:5).

While this invites the reader of *Something New* to do the opposite, and read the essays in order to find the genuine disconnection in his work, it also highlights the essay collection being, like *Tristram Shandy*, a book that anticipates the liberty of the immortal soul, which, as Richard puts it, "can at once look backward, before time was, and forward, when it shall be no more" (1:31). Richard's final aside to the reader in *Something New* makes this movement the basis of a telling analogy: "Is thought, said, I to myself, like matrimony, which preserves its knot intire, its yoke unbroken, under all the wanderings, estrangements, and alienations, of dissipation and prolifigency?" (7). He does not allude casually to marriage here. By the time he was writing these essays, Richard's marriage to Elizabeth was as settled as an unsettled reality could be. The two were still together, and had two children, born in 1752 and 1756, but their finances were precarious and many of their married years had been led in different parts of the country. Elizabeth, after working as a companion to a rich relative, had ended up as a playwright in London, and Richard, after trying and failing to make money from a linen manufacturing venture, spent much of his time avoiding the bailiffs.[50] Was marriage linear? Did it have a happy ending? If its chapters were as disassociated as those of his essays, and the free thoughts to which he lays claim, on what basis did its partners hold together? Richard and Elizabeth both seemed willing to answer such questions by suggesting that being married was not like living in a novel, but it was like reading a book.

As the mature writer of *Essays Addressed to Young Married Women* (1782), Elizabeth takes a hard look at the state of marriage in which women find themselves. Reflecting on the fact that women are generally uneducated about what to expect after a wedding, she emphasizes the fate of the long-suffering wife. "Let us suppose," she writes of a scenario that we can only assume was not so hypothetical, that the young wife's husband "has withdrawn his love from her and perhaps betrothed it on some unworthy object, to whom he devotes his time and fortune."[51] Under these circumstances, Elizabeth advises, a wife has no choice but to carry on. Her earlier spirit of outrage on behalf of women is still palpable—male behavior can be adjusted and scrutinized and reviewed, and her essays read as critical of men—but she makes it clear that marriage frames one's life rather than directing it or ensuring its outcomes. Like Sidney, Elizabeth works with a sense that things could have been different that arises from them being fixed.

This is also the message of *The Delicate Distress* (1769), Elizabeth's first and best-known novel, which claims to represent the "various contingencies of real life."[52] The novel was published as part of a set together with Richard's *The Gordian Knot*. The pair had written their novels in parallel, determined to have them appear together in a form that would capitalize on the reputation of *A Series of Genuine Letters*. Despite this collaborative constellation, *The Delicate Distress*, like *Amelia* and *Sidney Bidulph*, is thematically concerned with adultery and serial attachments rather than a single romance between its characters. Elizabeth's heroine and chief epistolary narrator, Emily Woodville, suspects her new husband of lasting affection for his first love. He confesses to this truth by writing, not to Emily, but to his male friend and correspondent. There is no way out of marriage, but readers are given every reason to understand it as anything but a happy ending. Although the situation is resolved in Emily's favor, *The Delicate Distress*, like *Amelia*, is unusual in making marital conflict and the realities of middle-class existence its focus.[53]

Elizabeth and Richard Griffith write, then, with a sensitivity to the ways even settled things go wrong and fail to line up: two does not always follow from one. While Richard's quirky texts are deliberately arranged to embody this principle, Elizabeth's novels, plays, and conduct book show women in situations of suffering and distress that postdate the moment at which other novelists would have resolved their fates by marrying them off. If marriage closes a novel, it nevertheless remains open in their work to interpretation and analysis. With this logic in mind, Elizabeth and Richard, who reject to different degrees the form of the novel as appropriate to the story of marriage they have to tell, embrace the form of the book. If marriage is not always happy, Elizabeth's *Essays Addressed to Young Married Women* suggests, it nevertheless has a format that enables revision and review over time. Richard suggests the same of his own pages of writing, which he presents as containing bursts of disassociated thought, pieces of writing that will hold together even as they are reread in any order we like.

This also goes some way toward explaining how Richard and Elizabeth approached the curating and editing of their own correspondence. For *A Series of Genuine Letters between Henry and Frances* is an unusual text. Despite the well-established conventions of the epistolary novel that might have compelled readers to follow the twists and turns of the relationship toward a happy marriage, the prefatory material makes it clear that these letters should not be read in this way. The happy ending, the editor points out, is to be found just pages away and may be immediately consulted: "it would not be amiss if the reader, before he proceeded, should turn over to letter cclxxv, where he will find the noblest and most rational arguments given for taking this step" (i:xx). Despite reference in the second edition to "this novel of our lives," Richard and Elizabeth present their letters as deliberately disordering the sequence of the correspondence, obfuscating connections between the original letters. Most letters are not dated. Whole sequences seem to be missing. Threads of arguments vehemently made in one letter are lost in the next. Some of this disorder, writes "Henry" in his preface to the second edition, is to be regretted: "We have processed to this second edition, complying with the preliminaries as far as we could do so with ingeniousness; correcting all former errors, presenting the true scene, restoring the names of places, filling up the blanks" (xxxiv). But the prefatory material added to the second edition also indicates that inclusion of new material has further disordered the letters' sequence: "There is not to be expected much Connection among the following letters, as they in Reality belong to the foregoing series; but having no Dates, were jostled out of their Places when I was endeavouring to frame the Suite from the Subjects of the letters themselves" (A4). These "several chasms in the series of our correspondence" have not been repaired, and the authors insist, again, on the naturalness of the compilation as genuinely disordered and incomplete. This supports Richard and the fictional editor's advice that we read them out of order. Richard also claims in his preface and in his own letters that his favorite way to refresh his appreciation of his wife's virtues is by reading a past letter of hers at random (i.vi). Of the published collection, he insists that he'd have thrown handfuls of the letters into the fire rather than have had them appear artificially ordered (i.viii).

In curating their letters, Richard and Elizabeth eschew the suggestion of their relationship being in any way linear. Instead the letters give the impression of a secure relationship, one that can be opened at almost any point, and read in any order, and reissued. Its outcome, like that of Booth and Amelia's marriage, is assured. But, with this in mind, random access is good, disordered reading is good, and disassociated readings are good. All of these modes of reading correspond to the state of marriage as Richard describes it, as one that "preserves its knot intire" despite wanderings and reversals. It confirms the acute sense that he and Elizabeth maintain in writing to each other for four years before their marriage that things might have been otherwise. As Wellbery suggests, being asked about one's partnership in modern terms is as likely to require an answer invoking contingency as fate. This analogy is as naturally accommodated by the interface of the book as it is excluded by the logic of the novel.

In *Impressions of Hume*, Davide Panagia makes an argument analogous to the one I have made here, for the material contingency attending the text's existence in time being something that informs its content. In his case study, this involves the reissuing of texts in new editions, which he links to Hume's politics of disassociation. Panagia argues that Hume's view of selfhood and of his own life involves a "commitment to contingency" that intersects with the technology of movable type—"a media technology that allows for the possibility of permanently shaping and reshaping, adjusting, editing, and revisiting parts so that new insertions of word, phrase, or image can be made here, extracted there, and conjoined in one way or another."[54] Hume's politics stays valuable in these terms because it "offers contemporary political theory a way of thinking the otherwise of givenness."[55] Hume's sceptical project, which throws everything from an object's continued existence in time, to the integrity of personal identity, open to question, becomes for Panagia a positive commitment to the certainty of the future's uncertainty. Hume's legacy frees us to think of existence as given in its parts, but not in its temporal unfolding. This is the aspect of publishing that allows Hume to relish the issuing of new "editions" of his own life. Such an outlook might allow all of us more generally to think of even the grimmest of any currently given scenario connecting up with better ones we cannot yet foresee.[56]

Richard and Elizabeth Griffith's view of books as objects open to multiple and uncertain futures can be read in similar terms. I have emphasized the codex book rather than movable type as the technology that comes into play in allowing multiple readings to inhabit the space of one text, but for the Griffiths it also matters that their letters can be edited. Like Elizabeth's fictional and literary representations of marriages that can be repaired, badly edited books find new life in new editions, and the authorized editions of *Genuine Letters* are distinguished by the different compilations of letters making up their introductions. The entity of the book holds together as it is corrected and reissued. Their approach to revising editions of their marriage becomes much like Hume's in revising his identity, with both suggesting that the present has a discontinuous relationship to the past and future. The Griffiths believe in their own letters, in their editions, and in their love for each other, and they believe that cause for hope lies in this potential for reordering, for reading, rather than in the prospect of finding any final or correct order. We might think here of William Rasch's description of Luhmann's systems in similar terms, as opening up "choice" through the temporal field of combination: "For an element to link onto another element, it must be able to distinguish among various elements and 'choose' one over the others—and this 'choice' must be made in time. This element A must link onto element X now, and not later, or element Y later and not now."[57] Richard and Elizabeth, who reject at many levels the causality and unidirectionality that the novel imposes on life, are strongly committed to romance in these terms at the unit of the page, and of the book as something that will be read and reread over time. As editors of their own letters, they are comfortable approaching their relationship as a collection of pages that allows past and future to be accessed and reordered from multiple points of view.

Elizabeth and Richard Griffith's experience as readers, writers, and editors complicates the idea of print reading, either as competing directly for everyday time, or as giving us a faster and more compelling version of linear time. The combination of fixity and openness that they perceive in the book, and the sense of contingency that this brings to scenes of its real and imagined reading, relies on one course of events being written down. But it also relies on time, not so much time to read, but time as the imagined vector along which unpredictability returns to books as they are opened and read in many constellations. Although online reading makes the dynamics of chance and possibility felt in its own way, it is the codex book that serves up in these terms the more specific awareness of contingency. What I read in print is given, in other words, in terms of narrative, but also by the covers of the book, the perimeters of my room, the suitcase in which I carry only certain things to read. Under these conditions, the way forward in time is open and every new combination of pages that I read flags others that might have been made and that contribute in this particular and unrealized form to the supply of alternatives. Desultory readers like Richard Griffith worked under these conditions with what they could lay their hands on, imagining less the deficit of time to read, or things to read, than horizons of possibility to which even a limited number of books can refer. If we want an antidote to the too-much-to-read thesis, it is here—in the idea that reading makes events that have been ordered one way into things that can still be accessed and reordered in time, and that therefore come with a surfeit of possibility that real life lacks.

CHAPTER 4

Time to Come

> Sometimes I think the books that affect us most are fantasy books. I don't mean books in the fantasy genre; I don't even mean the books we fantasize about writing but don't write. What I'm thinking of here are the books we know about—from their titles, from reading reviews, or hearing people talk about them—but haven't, over a period of time, actually read. Books that can therefore have a presence, or exert a pressure in our lives and thinking, that may have much or little to do with what's actually inside them.
>
> —Eve Kosofsky Sedgwick, *After Sex*

STOCKPILING

I lived in my later life as a graduate student in a university town with a great second-hand bookshop, one stocked by waves of books that arrived as their owners died or moved away. Composed of the books of bookish sorts, these collections were full of dignified editions, classics made to last. I grabbed Gibbon and Plato and Proust from the shelves. My quieter days of British student life by the sea were gone now, and I was studying in the United States, working harder on coursework and teaching than I'd ever done before. There was little time for reading lazily or randomly. But still I bought the used books keenly, with hunger for them as things I would come back to. After the jobs and the kids, there would be a small house on a hill, a big library, and a fireside. The books pile up and line my walls.

In a piece where he describes himself turning fortuitously to books that have sat unread for a long time on his shelves, Ross Posnock describes his own habit of buying books instinctively and leaving them unread over "a period of prolonged incubation." He calls this "deferred reading."[1] In the passage above, Sedgwick suggests that the point of actually reading may never have to arrive for such deferred reading to come into operation. Books in their unread state can still exert their influence. For John Durham Peters, "At times books you haven't read shine more brightly than those you have."[2] Knowing books in one way or another, referring to them, owning them, reading reviews of them, may be enough to secure their place in our lives and to shape our future.

This chapter is about the way books have made a quieter and more contemplative future seem graspable despite—or perhaps because of—suggesting that better times to come depend on a period for reading them that is still to come. We have already glimpsed several eighteenth-century readers hoping their individual circumstances will improve so as to allow them more time away from paid work (Thomas Turner, William Grenville), or from children, household, and parish duties (William Temple, Catherine Talbot). These hopes have been directed toward strategies of life organization rather than structural change. And on this count, Richard and Elizabeth Griffith, James Lackington, Elizabeth Carter, and Elizabeth Inchbald, to whom I now turn, really did find more time to read as they grew older and their finances grew less urgent. In Inchbald's case, her own retirement from the stage and focus on writing in later life converges with her fictional representation of reading as something difficult to reconcile personally with the demands of a woman's active youth. But it also resonates with a project of larger and more utopian dimensions, which I'll explore through William Godwin's writing: that of creating a future where leisure time would be more democratically distributed to all.

At the far horizon of this chapter—and this project more generally—is the larger question of how reading practices might intersect with the way we see work and its historical decline. In the 1790s, the serious, slowly digested book matters, I'm going to argue, to the idea that upcoming generations of workers will soon have more time for activities such as education, reading, and community building. From that decade until very recently, this was a common account of where history was going. Even late-twentieth-century statements predicting the decline of work and the increasing importance of leisure are common enough: in 1981, Ralf Dahrendorf, Director of the London School of Economics, wrote confidently of the "fundamental reduction of work in modern society" as an "irreversible" trend.[3] Since then, writers with different agendas, some predicting the need for us to live more quietly on a decimated planet, others foreseeing with some fear of social unrest the mechanization of the workplace, have continued to write of a future in which paid work will play a smaller role in defining human well-being.[4] But theirs have become minority voices in the twenty-first century: even as the need for workers continues to decrease, the idea of creating more paid work for more people has dominated recent ideas of political progress. Simultaneously, the idea of our needing more time to read has begun to seem less relevant in a world where text seems to flow around us in the most accessible of ways all the time and the prospect that we will read more in the future appears less certain.

The debates about Sunday described in the first chapter provide glimpses of an eighteenth-century struggle to limit working hours that was not yet formally organized. Texts like this one from 1804 extend the fight for a day free of labor in terms that anticipate the trade union movement on the horizon: "This benign institution is the grand bulwark of poverty against the encroachment of capital. The labouring classes sell their time. . . . Six days of the week are disposed of already. If Sunday were in the market, it would find purchasers too. The abolition of the Sabbath would, in truth, be equivalent to a sentence, adjudging to the rich the services of the poor, for life."[5]

By the 1790s, many writers were presenting secular visions of a future in which work would be better organized, and the time with books and for education would become a basic human right. As we've seen in previous chapters, the presence and availability of books and other reading material was instrumental in fueling the wish to give differently of one's time. The readers I've discussed so far suffer less from the feeling of the publishing industry producing too much to read than they do from the feeling that time to read what they have around them is undersupplied. But books in particular helped solve as well as create this problem by becoming conduits for a vision of a more leisurely time to come, one in which they would be accessed democratically and with pleasure. Books served in this role as objects of the kind José Munoz describes in *Cruising Utopia* as the "utopian bonds, affiliations, designs, and gestures that exist within the present moment."[6] For Muñoz, our orientation toward the as yet unrealized future is supported by real "objects and moments that burn with anticipation and promise."[7] In the later part of the eighteenth century, books, I want to suggest, began to smolder in just this way.

Writing in the 1760s, Talbot sets the scene for an ideal retreat in which her utopian life of reading might be possible: "Opposite the Entrance, are shelves filled with Books, of a serious and moral Nature, that take up one side of the Room. A Bed of plain white Dimity, with two chairs of the same, is opposite the Chimney, where a cheerful Wood Fire is continually blazing. Near the Fire is placed a little Table, and a low seat, more for convenience, than show, and the Walls are covered with white Paper, over which a Vine seems to spread its leafy shade."[8] The books here are physical objects, the only things not downplayed as "little" or "low" or "plain" in this simple room. They are hefty in quality and serious in nature. They don't invite reading of the desultory kind Richard Griffith likes. But nor do they overwhelm the setting with their demand for attention. Instead, time appears on the side (the fire never goes out). By the 1790s, the writers I turn to here were invested in this fantasy on a larger scale. By now the difference at stake was not only between Sunday and the working week, but between youth and maturity, and between a present and a better future. Talbot's study becomes in this light a depiction of what a happily quieter age might look like—not as an afterlife, or a stream of Sundays, but as a period of history to come when everyone, especially women and workers, would have a chance to tune in properly to those books already at hand.

The materiality of books, both as things that visibly exist before they are read, and as things that can legitimately be read later, comes into play in the account I want to give of the way the future was imagined in the 1790s. Understanding them in this way will involve some readjustment of the more common angles at which to look at their physical features. For the art of seeing bindings or imprints or past uses for books is most commonly connected with efforts to understand history. To think about how texts were made, printed, edited, and read is, as D. F. McKenzie stressed, one way to access the human history to which their texts speak.[9] A new emphasis on comparative media has in some ways changed this orientation by making the physical questions of storing, accessing, and reproducing texts ones directed toward the future. But many of those interested in digital media remain concerned primarily with the problem of our access to the past. For instance, Jerome McGann's case for the relevance of philology to this digital age emphasizes the role of media and book historians as custodians of the past, scholars tasked with the ability to "preserve, monitor, investigate, and augment our cultural inheritance, including all the material means by which it has been realized and transmitted."[10] Wolfgang Ernst has a similar eye to the problem of memory when he sets up books as the "first external memory devices through which culture as memory based has been made possible," suggesting that paper and print have been replaced by other inscription techniques that are also concerned primarily with memory.[11]

In both practical and theoretical terms, then, it is much harder to talk of the future being propped open by the book's physical form than to talk of the way books have brought the past into being. Bernard Stiegler, whose *Technics and Time* makes one of the more ambitious but less media-specific cases for all our experiences of time being mediated by technology, asserts with relative ease that historicity depends on writing: it is writing that allows us to record, forget, and rediscover experience as the past.[12] Technology is also easily implicated in recognizing the self-identical time of the present: I need a clock, for instance, to tell me what time it is now. But what about the future? Why do I need technology to conceive of what is to come? Why would this involve writing? The challenges of Stiegler's argument really begin when he addresses the text as original to our conception of time, not only as the past, but as a future still to come. He draws here on Derrida, for whom the archive anticipates the future, while emphasizing that the future anticipated through technology differs from the one Heidegger has in mind as our orientation toward death.[13] Instead, Stiegler argues, writing introduces a form of deferral that we inhabit: "A being who differs and defers should be understood in a twofold sense: the one who always puts off until later, who is essentially projected in deferral, and the one who, for the same reason, finds itself originarily different, indeterminate, improbable. The being who defers by putting off till later anticipates: to anticipate always means to defer."[14] Without stressing that this makes book reading something that never comes, or the figure of the reader one who never arrives, Stiegler's concept of deferral concretely engages writing's materiality in a way that might allow us to see the codex book extending time rather than anticipating finitude. By becoming a prop in the imaginary drama of what will come, books advance a future that can be weighed up (held, shelved, placed by the bedside, talked about, and quoted) even if the time it actually takes to read them remains deferred as a perceived difference between now and the future.

This resonates with the future as Luhmann defines it, as a time that does not exist, and cannot arrive, except as the relation between what is past and what is to come. This difference can be told differently at different times. Today, for instance, increments of temperature increase support the way we see into the future; in the 1970s, Luhmann suggests, it became visible as the calculation of risk.[15] But in the 1790s, I want to argue, it could be felt distinctly as the difference between the book, given in its tactile wholeness as an object, and the book as something readers would come to spend time with in the future. That difference becomes part of the constitution of the book as distinct from other media (plays, pamphlets, speeches, newspapers), which were and remain much harder than books to have in advance of their reading, and much less visible as objects that can be stored for later. In today's terms, we are routinely invited to download files to "read later," but it's harder to speak of unopened files operating in the way that Sedgwick and Peters and Posnock speak of unread books, as active constituents of reality, active agents of temporal differentiation, in that time before they are read.

ROMANTIC MEDIA

In some ways it is, of course, counterintuitive to claim that books featured in the revolutionary imagination of the late eighteenth century. The more obviously new and phenomenal media at this time were periodical pieces, squibs, and poems, texts that could move cheaply and be read quickly, or the speeches, plays, and caricature, which resonated more widely and with more quickly formed publics than books. The flowering of pamphlets, an accelerated speed of print transmission, the growth and regularity of postal networks, and a multiplying of literary forms matter much more to that sense of urgency, quantity, and periodicity that critics including Clifford Siskin, William Warner, Michelle Levy, Mary Favret, and others have associated with media ecology of late-eighteenth-century culture.[16] Paine's first edition of *The Rights of Man*, to give just one example, was quickly and cheaply published in two parts in 1791 and 1792 and made an impact largely because of its pamphlet form.[17]

In contrast to these fast-flowing media, books were widely seen as antiquated and unfashionable, even by their own late-eighteenth-century purveyors. Vicesimus Knox introduces his collection of short essays, *Winter Evenings: Or Lubrications on Life and Letters* (1788), to a world overflowing with new things to read. Apologetically, he suggests, "as my volumes are not very large, there may perhaps be found a little crevice in the world, into which, provided you are really gentle and good natured, you may find means to squeeze them."[18] Here, books appeal by making themselves small and meant for occasional consumption, strategies we've seen in play in earlier chapters. But other approaches to books at this time suggest that their larger scale and permanence might be a relief to the reader. Books, or pamphlets made into books, made reading easier to defer than newspapers or letters or broadsheets. In the satirical *Miseries of Human Life* (1806), the narrator complains of the problem of having "pamphlets and loose printed sheets daily getting ahead, and running mountain high upon your shelves." His solution is not to read them, or discard them, but to "tame them by sorting and sending them to the binder," producing in effect a book that becomes an archive for the future.[19]

In his *Memoirs* of 1791, Lackington explains that books were unlikely to be bought or consumed by a public hungry for news. Their advantage was that they could be consumed when other kinds of reading and entertainment went quiet:

> I have always found that bookselling is much affected by the political state of affairs. For as mankind are in search of amusement, they often embrace the first that offers; so that if there is anything in the newspapers of consequence, that draws many to the coffee-house, where they chat away the evenings, instead of visiting the shops of booksellers (as they ought to do, no doubt) or reading at home. The best time for bookselling is when there is no kind of news stirring; then many of those who for months would have done nothing but talk of war

or peace, resolutions, and counter-revolutions &c &c for want of other amusements, will have recourse to books.[20]

Lackington's assessment suggests what systems theory would describe as two different systems coming into play, in both of which reading plays a part: a media system that depends on speed of consumption, and an art system that depends, as we saw in the case of Sunday reading, on a slower and sectioned-off form of engagement with texts. These two systems sort texts on the basis of reading speeds and times as well as genre, but are nevertheless in competition in Lackington's account when it comes to how one reader is to spend an evening. The evenings in which working people might retire to read books are given over to faster kinds of textual engagement and conversation as soon as current affairs become compelling. In Lackington's view, and in Knox's, books get picked up only in those hours when the present is quiet enough. This prognosis suggests why books, especially old books, might be seen as of limited use in any revolutionary climate.

"The Adventures of a Robinson Crusoe" (1799), a quip on the novel's redundancy published in *The Young Gentleman's and Lady's Magazine*, suggests the same limitation. "I have been of some utility in my day," claims the copy of *Robinson Crusoe* that serves as the inanimate narrator of the piece, "though the magazine's novelty has now supplanted me."[21] In the next four pages, the speaking book describes itself as having been carried home a generation earlier from a bookshop by a young lady, who read it constantly and with great excitement. But by the time this young woman grows up, books are out of fashion. She gives her copy of *Robinson Crusoe* to her son, where it lies among the "literary lumber" of his school books before being passed around among his friends, sold for baked apples, thrown into a deal made in exchange for a copy of *The Young Gentleman's and Lady's Magazine*, and finally eyed as scrap paper by the chandler. Eventually its physical downfall becomes complete, explaining the novel's reappearance as the paper substratum of the magazine piece we are reading. Godwin and Inchbald, as we'll see, subscribe to these concerns about the redundancy, elitism, and slowness of book reading, and both of them recognize the advantages of other forms of writing, particularly for the stage, over the book.

And yet it's also precisely in this context that books surface, not just with the patina of old objects, but with the glow of prophetic ones. There are two different dimensions in which to describe their role in boosting the fantasy of leisure to come, both of which become evident in the arguments Godwin makes for the slow and deliberate reading of unfashionable books. In the first place, as physical objects books appear generally in contrast to serial publications and oral performances, as sequences of text where an ending is in sight before it is read. "Wherever we are in the reading of a written narrative," writes Mark Currie, "we have access to the future, to what lies ahead of us in the discourse, in the sense that it is there to the right."[22] Stressing the way in which the book's block view of time differs from the experience we have of time in real life, Currie elaborates in his reading of twenty-first-century novels on the juxtaposition of the untensed time of the book (where past, present, and future all exist at once) and the progress through time that is entailed in reading a text. From the perspective of the book itself, Currie argues, "the future exists, and the ontological priority of the present is an error produced by the more psychological experience of time."[23] In material terms, a book is already made, but from a potential reader's point of view, it is still to come. These options make no difference to its constitution. Tense, Currie concludes, is reintroduced to fiction as the vector along which one reads, but it is not native to the book as form.[24]

Currie, whose interests are largely narratological, does not develop the consequences of this argument in medial terms: What happens, for instance, when a novel is serialized, as so many in the nineteenth century were? How does this apply to an audiobook, which is much harder just to flip to the end of? One cannot easily turn when listening to a cassette, or even to an audiofile, to the end of the book as something already there, at least not with the casual gesture that many readers report using to preview a novel's future.[25] But even as he assumes that the native context for the novel is the book, Currie locates the tension between the book's already-thereness, on the one hand, and the sequential progress of the reader, on the other, in ways that are useful in thinking about the views Inchbald and Godwin hold. Just at the point where pamphlets, squibs, caricature, and plays appear as media formats with a rapid turnaround, texts written and published as books become palpable as objects in which the future can be fully unfolded *and* yet still in wait, both strangely graspable as possessions and yet distant as objects belonging to a far-off and utopian future in which they will have been read.[26]

The other dimension books bring as media to the time of reading involves their durability and the idea already mentioned in relation to Knox and Lackington and Talbot—that they can wait for a quiet season to come, a turning to books that happens once theater, news, work, and the social exertions of summer have died down. Temple, whom we met in chapter one, definitely sees winter in these terms as a time for reading books. This takes us back to Kittler's point about the nature of all writing, as a recording technique that lags behind the reality that it translates into signage. But there's a difference Kittler overlooks, upon which Knox, Temple, and Lackington rely in their representation of books stockpiled for later. To newspaper reports or text, such a lag is lethal. To books it is natural. The time of their publication and their reading diverges much more naturally than in most other forms of print. And the more surely a book is written and made to last, the more positively that distance between the time of its being produced and that of the reader's time is expanded and affirmed. As such, the book brings the time of reading into focus as something split off from daily life, setting it up as likely to be out of kilter with the self-identical time of daily life, of clock time. If the book's time is grasped partly as the time in which it will have been read, its materiality supports a difference between now and that future. In its objecthood, visible against the backdrop of other more immediately compelling media forms in the late eighteenth century, the book reckons in increasingly concrete and upbeat ways with this difference.

A SIMPLE STORY: READING COMES LATER

It is true that I have always thought of reading as something I'd do more of later. It's true, too, that many people I know prove such an aim possible. I have friends and relatives who read long and deep in their retirement, finally making their way through all of Trollope or Gaskell and new translations of Tolstoy, many becoming the best of all imaginable interlocutors about literary things. My father, who did not finish high school, has been a regular at the local library for years, living in a little shack by the sea that is falling down around his ears, replenishing his fridge with white wine and his shelves with books in equal measure. He told me recently he'd just read everything by Naomi Klein, an author I'd been cajoling undergraduates to read at the rate of thirty pages a week. But it's also true that most of us will probably never read as many books as we did in our youth. This is unlikely to be the direction in which history is moving us, toward futures of more book reading. I find this a cold, hard truth, an upending of a life strategy I once held secure.

Yet the life of Elizabeth Inchbald (1753–1821) did involve more reading as she grew older and her professional life of work and travel abated. She began her stage career early—incidentally, with an application to the theater manager Richard Griffith, brother of the Elizabeth Griffith discussed in the last chapter.[27] When this application failed, she went to London and married the actor Joseph Inchbald. Throughout Elizabeth's late teens and twenties the couple toured the United Kingdom as actors, often accompanied by their friends, the theatrical siblings John Kemble and Sarah Siddons, and by Joseph's son from a previous relationship, who worked as a child actor. The Inchbalds spent time in Scotland and Ireland, traveled to France in the hope of a quieter life as artists, and returned penniless to England to take up another string of low-profile theatrical engagements. Their life was a strenuous and social stream of rehearsals and learning of lines: in a single month when the Inchbalds were lodged in Aberdeen, Elizabeth played Juliet in *Romeo and Juliet*, a witch in *Macbeth*, Cordelia in *King Lear*, and Clarissa in *All in the Wrong*.[28]

Under these conditions, it is hardly surprising that Elizabeth read little more than scripts. The Inchbalds rarely spent more than a few months in one lodging.[29] Although her notebooks from the 1770s indicate that Elizabeth had plenty of access to literature and that she and her friends often read to each other, she didn't own books. Most of her time seems to have been spent in marital conflict, travel, and arduous preparation for performances in which she rarely excelled or felt confident. In this context, her reading, much of which focused on her learning French, was sandwiched tightly between work duties. On May 26, 1777, for instance, Elizabeth rises at three in the morning and leaves Manchester with her husband, Kemble, and Siddons. The four arrive at Birmingham, where Joseph, an aspiring artist, paints in Siddons's apartment while Elizabeth goes through her parts with Kemble. Her reading for the day happens late and consists of a little *Telemachus* in French and an abridgement, also in French, of the Bible.[30]

In 1779 Joseph died suddenly, leaving Elizabeth in mourning and concerned for her own financial security. Although she continued to act, to tour, and to lead a highly social life, attracting several new suitors, she now began to prioritize her own writing. By the end of the 1780s, she had fended off prospects of remarriage and her plays were earning high fees. Her journals from this decade show her writing steadily between rehearsals, visits, and moves, often while depressed. But her 1780s reading, primarily of newspapers and new productions, seems to happen mostly when she is having her hair dressed or in snatches that leave no obvious trace on her thinking. Typical, for instance, is a "dullish" day in 1781 when she reports: "Mrs. Hunt read a story to me while I dresst__Mr Webb came home to dinner, brought two newspapers, where I saw the Baron by Mr Andres was condemned__began the Life of David__walked__wrote but did not please myself."[31] Here, from the same period, is a similar day: "Laid in bed very late and was melancholy, heard Miss Mills had taken the dining room in the afternoon, read a little out of many books, drank coffee in the kitchen, it thundered the time—after at Matilda till dark then Mrs Hitchcok just called who came with Mrs Mills—went for letters but there were none."[32] In this context, Elizabeth Inchbald manages to write, producing what seems from her description here to be an early draft of *A Simple Story*. Yet she finds it hard to justify spending any length of time with a text not her own and most of her reading experiences are subsumed to a wider range of interactions that compete for her attention.

A new calm came to her life in 1790, however. Beginning the year ill, but no longer worried about money, she turned to her desk and to solid hours of study. Retired from the stage, she read Rousseau's *Confessions* and Johnson's *Rasselas* and began to work in long, concentrated bouts on *A Simple Story,* the draft of which she'd abandoned nearly a decade earlier. She rented one relatively commodious London apartment and socialized much more on her own terms. While news of the world flooded England, Inchbald tuned out of contemporary reading and into older texts, relying increasingly on others for news.[33] Like Lackington, who found himself reading at the end of the century the books he'd been handling all his life, Inchbald read in her late thirties and early forties the books that shape what we now think of as her youthful voice of the 1780s. Her slower and more dedicated months of reading fed back into writing a novel, an undertaking not at all to her taste. She wrote to Godwin at the end of 1792, "I do not shrink from labour, but I shrink from ill-health, low spirits, disappointment, and a long train of evils which attend laborious literary work. I was ten months unceasingly, finishing my novel, notwithstanding the plan (such as you saw it) was formed, and many pages written."[34] She had, she went on in this letter, "frequently obtained more pecuniary advantage by ten day's labour in the dramatic way than at the labour of this ten months." Inchbald is conflicted about the advantages of writing at this pace, but her determination to continue with it underscores the more positive view she held of books, particularly in relation to theater, as a medium in which the passage of time could be accounted for in more complex ways, and anticipated as being on the book's side when it came to the question of long-term reception and posterity.

This chronology applies to the interpretation I want to give of *A Simple Story* as a fiction that explores the time of book reading being pushed out chronologically to the limits of the story to which it also becomes integral. The novel, published in its final form in 1791, remains the best known of Inchbald's works and the more celebrated of her two novels. It is widely beloved for the feisty character of Miss Milner, the independent young heiress who dominates the first portion of the novel, and it is still praised, most famously by Terry Castle, for the transgressive energies it unleashes on women's behalf.[35] As the novel opens, Miss Milner comes to live with Dorriforth, a Catholic priest who is her guardian—and soon to be her lover and eventual husband. Together with their respective companions, Miss Woodley and Sandford, Miss Milner and Dorriforth make up a household saturated in tension, conversation, and desire. The love between Miss Milner and Dorriforth is not easily declared or resolved: as a novelist, Inchbald offers sparse psychological detail. Yet her characters share a house throughout their courtship and after their engagement, allowing Inchbald to portray their attraction in highly theatrical terms, through gesture and dialogue. This domestic setting is unusually gratifying to read about, perhaps because her energetic, dialogue-driven, and well-populated scenes capture aspects of the life that the Inchbalds shared with Siddons and Kemble. The first two volumes of *A Simple Story* provide, in short, a realm of quarrelsome but gratifying interaction that gives women a role as agents of their own desires that others novels of the period deny them.

The second two books of *A Simple Story* are set seventeen years later in a much quieter period of time. At this point Dorriforth, now Lord Elmwood, and Miss Milner have married and separated, and Miss Milner, now morphed into the adulterous and much chastened Lady Elmwood, has died in exile on the margins of civilized society. The daughter, Matilda, who is left behind, is Dorriforth's legitimate heir, and her mother pleads protection for her in a last letter to Dorriforth. This sets up the second half of *A Simple Story*, in which Matilda, admitted to her father's house on the condition that she does not see or speak with him, spends her time without visitors and in the specific company of books. Lacking the verbal wit and spirit of her mother, imprisoned within rooms where her mother once had free rein, Matilda becomes the heroine of a fairly trite and predictable series of plot twists that secure her reunion with her father and her own marriage without her personally having much to do with the outcome.

All evidence suggests that Inchbald began, and in fact tried to publish, the novel when she was still a young and busy actress, and that she completed its second part and revised its first much later, in 1790, as she retreated from that life.[36] The details of just how different Inchbald's life looked at these two points in time sheds significant light on the quite different roles that reading plays in the two parts of *A Simple Story*. While the first part of the novel is packed with dialogue and action that account for most hours of the characters' days, the second reneges on the appeal of that all-consuming, social present in favor of reading. In the first part, Miss Milner shops, she goes to balls, she entertains her various lovers, she argues incessantly with Sandford, and she performs in outspoken conversations with Dorriforth. These scenes of action reduce books mostly to props. Dorriforth, for instance, pretends to read as he hides his anger over Miss Milner's trying behavior.[37] Later, after their love is declared and while it's being put to the test, she comes home having purchased books:

> She had been passing the whole morning at an auction, and had laid out near two hundred pounds in different things she had no one use for; Among the rest was a lot of books on chemistry, and some Latin authors.

> "Why, madam, cried Sandford, looking over the catalogue, where her purchases were marked by a pencil, "do you know what you have done? You can't read a word of these books."
>
> "Cant' I, Mr. Sandford? But I assure you, you will be vastly pleased with them when you see how elegantly they are bound" (146).

Miss Milner's comment is pitched as provocation but it suggests with some seriousness that her pace of life necessitates her seeing *most* books as objects of display rather than consumption. Although Dorriforth urges her to slow down, pressing on her "the necessity of time not always passed in society; of reflection; of reading; of thoughts for a future state; and of virtue acquired to make old age supportable," she shows no inclination to follow his advice (44). Miss Milner's lack of interest in study, in contemplating the future, is undoubtedly a failing. But it also corresponds to the way Inchbald herself seems to have felt during the years she was inventing this character, where hours spent working and socializing subtracted directly from the time she might have spent reading or planning for her own future.

The force of Dorriforth's corrective toward a deeper kind of study and contemplation is strongly felt by the end of *A Simple Story*, which ends notoriously by mourning Miss Milner's lack of "A PROPER EDUCATION." The education Inchbald has in mind for Matilda seems to involve, above all, spending time with books that Miss Milner had contact with but did not open. As a reader, Matilda becomes the mirror image of her mother on this count: "Educated in the school of adversity, and inured to retirement from her infancy, she had acquired a taste for all those amusements which a reclusive life affords—She was fond of walking and riding—was accomplished in the arts of music and drawing, by the most careful instructions of her mother—and as a scholar she excelled most of her sex" (221). These tastes serve her well during the months she is confined within her father's house. Books, decorative objects in the first part of *A Simple Story*, are now key actors in its plot. The unread (and unreadable) volumes that had been sources of contention in Dorriforth's relationship with Miss Milner, become crucial points of contact in relation to his daughter. During Dorriforth's absence from the house, Matilda looks "with the most curious attention" at the books on his reading desk (245). And when he is there, Dorriforth / Lord Elmwood chooses with a meticulous but invisible hand the items to be sent to her from his library, sending a pile of his selection back to her quarters (265). Matilda uses the gardens when he is away, but during his periods of residence she studies, sensing that the passage through pages is the only route into her father's life.

With this chronology, one in which a period of engagement with books follows and preempts an intense period of action, *A Simple Story* introduces a disjuncture that many critics have found unsettling.[38] But this split can also be accounted for by saying that the second part of *A Simple Story* explores reading as a form of time use more palpably than most novels. Matilda's reading, for instance, becomes obvious as a way to *do* rather than to forget time. At Elmwood Castle she devotes "certain hours" of each day to study with Sandford and others to music and riding (221). And Inchbald herself constructs, at least in her second period of its composition, a novel that frustrates in similar ways our compulsion to move forward. Omens of unhappiness, most famously the mourning ring with which Miss Milner and Dorriforth are united, make reading for the plot feel inadvisable at the novel's every turn. Rather than being urged to rush toward an ending that will bring lovers together, we are deterred from this axis of movement by *A Simple Story's* having already witnessed an exhilarating and fully blown state of attraction between its main characters. Marriage, with which an eighteenth-century novel often ends, could hardly better the intimate conversation in which the first generation of lovers are here involved from the outset: as a spatial arrangement, this book feels already all there.

And yet time, which reading seems to take more of when it is not fueled by hope of a happy ending, has still to pass. It does so palpably in the second part of *A Simple Story*, where the weeks between March, when Matilda arrives at Elmwood Castle, and August, when her father arrives, pass "in peace, content, though not in happiness" (221). When Dorriforth takes up residence late in the summer, Matilda lives "within a few rooms" of him, in an arrangement that recalls the first part of *A Simple Story* by again making proximity the desired outcome and the status quo of the story. The books that traffic between the rooms of the partitioned house provide no shortcut into his space. Matilda cannot have these books without their time of reading; she cannot skip pages. In this way, Matilda's portion of the narrative introduces duration and chronology to a tale that originally spurned it by being all action. If, as Emily Anderson puts it, the first part of *A Simple Story* gives us a series of characters "trapped within performances," playing parts in a novel indebted to the theater for its pace and level of action, the second part gives us a story and characters that feel recycled from fiction, and digestible only as reading.[39]

Mikhail Bakhtin's idea of the chronotope is famous for fusing time and space, making it possible to describe roads and houses in fiction as zones as temporal in narratological terms as they are spatial. But when he gestures to the creative work having its own chronotope, Bakhtin stresses a much looser conjunction, with the narrative contained in the work being overlaid by the event of its reception: "These events take place in different times (which are marked by different durations as well) and in different places, but at the same time these two events are indissolubly united in a single but complex event that we might call the work in the totality of all its events, including the external material givenness of the work, and its text, and the world represented in the text, and the author-creator and the listener or reader."[40] Inchbald, I'm suggesting, thinks in similar terms of the space of the book and the time of reading as conjoined elements that can be disarticulated. Through education, she introduces duration to an equation that seemed at first as if it could be spatially represented: Matilda's time of reading is imposed retroactively upon her mother's space of being in the house. Miss Milner's happiness predates the passage of time that should in narrative terms have led up to it. And we, who participated in Miss Milner's impatience, experience this wait ourselves by reading chapters devoid of Inchbald's witty dramatic dialogue and lively scene-making. Rather than being the source of instant gratification that narrative was in the first part of the novel, in the second part our page-turning becomes a drag, a deferral of the revolutionary events and conditions that Inchbald had already brought to life.

This might suggest that there is a punitive side to Inchbald's revisioning of her earlier scenes, a regret in her own life at not having slowed down earlier to accommodate reading. She seems chastened for her precipitation in granting happiness to characters who have not yet done their time in the prison house of narrative tribulation. It might be argued that Inchbald makes reading part of what Soni assumes to be the novel's trial, foisting books upon Matilda as objects with which time must be spent if she is to redeem her mother's story. This would accord with Jane Spencer's reading of the second half of the novel as a "kind of atonement on Inchbald's part for the boldness of the first."[41] And it would resonate with the case Marcie Frank makes for Inchbald's late vote of confidence in the novel as a genre that allows one to make a better political case than one does on stage, *because* of the lengths of time only a novel can represent.[42]

Yet Inchbald also subverts what Soni assumes to be the only chronology open to the reader, one that involves moving forward through a book from left to right toward the promise of happiness. In *A Simple Story*, as with other novels I've mentioned (*Sidney Bidulph*, Austen's novels) all the components of happiness are there at the beginning. This makes that forward movement though the book a measure of time disconnected from the characters' spatial advance toward happiness: one could, as Austen registers, just skip to the last page one already has in hand. For Bakhtin, this describes a chronotope in which the creative work is first finished and *then* consumed. It also recalls Currie's description of the physical book that lies there spatially in its untensed form before it is unfolded incrementally as reading. With Inchbald, this chronotope, normally so different from the merging of space and time within narrative, becomes intrinsic to her plot: one might first have happiness and only later find the time to read about it.

In literary terms, this could become a case for *A Simple Story* as more unified than is generally thought, with its first installment deliberately resituated in time by its second. But in historical terms, it also takes us back to Inchbald's own life. Having begun to write *A Simple Story* in the midst of her captivatingly crowded days on stage, the mature Inchbald finished it with more time on her hands. James Boaden, her nineteenth-century biographer, and Isobel Jenkins, her more recent one, both describe her fairly abrupt cessation of her lively friendships of the 1780s with Holcroft, Wollstonecraft, and Godwin, and her return to Catholicism. As in Opie's case, Inchbald's withdrawal from those radical circles becomes in her biographies evidence that she was never quite at home in them in the first place.[43] But a more practical and progressive logic is evident too. As a result of her financial prudence, skills as a writer, and objection to marriage, Inchbald was able to carve out for herself three decades of literary engagement during which she seems to have been grateful to be out of the public spotlight of the theater. While she clearly works hard as a writer, the need to legitimize writing as a form of work, something that has been discussed in the case of other women writing at this period by Jennie Batchelor, is lessened in her case because she has earned her reputation in another arena.[44] Yet Inchbald's turning the wattage down on her earlier social life need not suggest that she is tuning out of the conversations and withdrawing from networks where real change happened in the 1790s. By settling down to read and write after her wildly social youth, Inchbald can also be seen as reaping in a single lifetime the benefits of protest that it takes her fictional characters two generations to realize. In this sense, *A Simple Story* participates fully in the logic of a better time to come that Godwin held to be most revolutionary, and to which I will turn in the rest of this chapter.

A Simple Story's coordinates are also compatible with Godwin's view of social revolution as both the prerequisite *and* the outcome of education. While Miss Milner / Lady Elmwood and Matilda's spaces of exile in *A Simple Story* are gothic and terrible, dramatically out of step with the contemporary, the contemplative margin they restore to the whole novel is not just one of punishment: it is one of time for literary consumption upon which the whole novel's chronotope depends. While we have arrived at—and can get no further than—the happiness of the young and independent woman with which *A Simple Story* began, Inchbald, by withdrawing from and returning to that scene at a contemplative distance, aligns herself with the more radical arguments for a post-revolutionary future, one in which we'd all sit down to read. She uses the novel to give women's revolution a time frame that it struggles at a daily level to supply itself.

GODWIN: THE FUTURE IS NOW

Inchbald's sometime friend, William Godwin, offers a more politically worked out version of this chronology, whereby reading is imagined as coming after the book and after the change on which its reading proleptically relies. In Godwin's case, evidence for this future exists partly in the connection he weaves, in his own life and in his writing, between the time given to the reading of old books and the democratic prospects that he envisions for the wider population of British workers. If Inchbald senses the problem of the protagonist who wants to act and to read, and makes a covertly political point by suggesting that it takes two women to live that one ideal life of both actor and reader, Godwin homes in on the problem more directly by flagging the fact that for most people, leisurely access to books is restricted. In the present, he states in his *An Enquiry Concerning Political Justice* (1793): "Literature, and particularly that literature by which prejudice is superseded, and the mind strung to a firmer tone, exists only as a portion of the few. The multitude, at least in the present state of human society, cannot partake of its illuminations. For that purpose it would be necessary, that the general system of policy should become favourable, that every individual should have leisure for reasoning and reflection."[45] Throughout his argument, Godwin stresses that "the poor are kept in ignorance by the want of leisure," with the result that "at present, ninety-nine persons in an hundred are no more excited to any regular exertions of general and curious thought, than the brutes themselves" (423–24). It is worth recollecting here that he's right: as E. P. Thompson and others have already told us, for most people, working hours increased sharply in the last part of the eighteenth century. Godwin is therefore identifying a conundrum we've seen played out in different ways throughout this book, namely that all arguments about increased quantities of reading in this period need to be offset against the sociohistorical fact that working people were losing the leisure time in which to pick up those books. A future such as the one Inchbald made for herself and her characters through withdrawal from social duty and theatrical performance is grasped in this context by Godwin as a political ambition, a spur to imagining democracy as a form of government that would depend for its operation on the general increase of "leisure for reasoning and reflection" (22).

Yet Godwin's reading in his own lifetime was steady and impressive by any count; his contact with books was regular, avid, and wide-ranging. In his unfinished autobiography, he notes rising as a student at five in the morning and reading until midnight.[46] On a single and not entirely unusual day, September 12, 1791, he records in his journals that he has read part of Hume's *Morals*, Paley's *Principles of Moral and Political Philosophy*, Samuel Johnson's pamphlet, "Taxation no Tyranny," and James McIntosh's *Vindia Gallicae*.[47] This was during a period in which he was also working diligently on his *Political Justice*, a text into which most of his reading fed. But once *Political Justice* was in press, Godwin's book consumption became even more impressive: in 1793, the year it was published, the recent editors of his diary note that he mentions reading 892 times, a spike in engagement with literature that suggests he was studying just as extensively once his attention had moved to *Caleb Williams*, the novel he published in 1794.[48] During his life Godwin had, Lamb reportedly gibed, read more books not worth reading than any man in history.[49] In relation to the other readers I've described in this study, Godwin was probably the most prolific.

Godwin was often an unfashionable reader, following his own instincts in his choice of books, reading things long after they had been published and allowing the speed at which he read to be dictated by his own habits rather than the press. His book collection was described as irreplaceable.[50] Educated as a Dissenting minister, Godwin read both Greek and Latin and returned routinely to classical texts. In the early 1790s, Godwin, like many others, read Voltaire, Rousseau, Hume, and Montesquieu. He read a draft of Inchbald's *Simple Story* before its publication. But alongside these authors, he read Nicholas Rowe, Richard Steele, and Thomas Otway.[51] Such reading practices would have made Godwin the ideal customer of bookshops like Lackington's, a space where old and new books crossed paths according to the deals Lackington could cut rather than their current popularity.[52]

In *Advice to a Young American* (1811) Godwin explains his own preference in books, advocating, above all, the state of mind that comes with diverging in one's reading from the texts fashion dictates. Soon enough, he argues, the reader must "be plunged into the more sordid realities. . . . I could wish that those who can afford the leisure of education, should begin with acquiring something more generous and elevated."[53] Selecting the best books involves finding the ones of a certain vintage and tone. "A young person," argues Godwin, "is to be very moderate in his attention to new books. In all the world I think there is scarcely anything more despicable than the man that confines his reading to the publications of the day: he is next in rank to the boarding-school miss who devours every novel that is spawned forth from the press of the season."[54]

In general, essays like "Of Choice of Reading" in his 1797 *Enquirer* suggest that Godwin can be quite agnostic about the content of books, assuming that a good reader will be at liberty to choose from a wide range of material. A child, he argues here, should be allowed to find books in his own time, and to "wander in the wilds of literature."[55] But he is prescriptive about the frame of mind, the "temper" in which the best reading happens, and the "true mode of reading"—one that requires leisure and reflection. For Godwin, a little like Sir Charles Grandison, the ideal state in which to read is at a distance from daily life, rather than in obedience to it. Put in terms of *A Simple Story*, this means reading more as Matilda does and less as Inchbald herself did as a young actress.

Although he consumed books in huge quantities his whole life, by the 1790s Godwin was no longer reading nineteen hours a day. His time for reading old books was in the morning before breakfast, and he set aside those early hours deliberately for that purpose. Reading before turning to writing and the events of the day allowed him to do so in the temper he advocated so strongly as a measure of the best reading. But, as the modern editors of Godwin's diaries state, and as Mark Philp has used these diaries to show, the largest part of his time in the 1790s was not spent at his desk. In the evenings and afternoons, he joined the conversations of which he was wary in print, and he went avidly to the theater. His diaries from this period show him making and receiving calls at an impressive rate; corresponding widely, often with romantic intent; and participating in the Philomaths, a society that met on Tuesday evenings to discuss topics of the day. With the publication of *Political Justice*, his social circles expanded significantly, making his conversational life all the more time consuming.[56] While there are different perspectives on just how keen Godwin was to participate in or speak to the more inclusive public conversations of his day, it is clear that he valued the contrast between his unplugged morning hours and his life of social contact and conversation during the rest of his day.[57]

This division of time, between books and conversation, was one Godwin took up in theory as he described the ideal way in which democratic ideas would spread and lives would be led in the future. In the utopian state he describes in the sixth book of *Political Justice*, everyone will do a small quota of manual work, but everyone will also have the opportunity to study in private and to participate in enriched forms of conversation. "Is there not a state of society practicable," he asks in the *Enquirer*, "in which leisure shall be made the inheritance of every one of its members?"[58] Godwin was not the first or the last to imagine this ideal day of the future, or to perceive the fact that if work were more fairly distributed, more time would become available for leisurely intellectual cultivation. The female inhabitants of Sarah Scott's *Millennium Hall* (1762) work limited hours that leave them largely free for festivity and the cultivation of the arts. The very structure of the novel conveys this pace of life, as the history of the institution and its inhabitants is told over several days to incredulous male visitors, for whom the dreamlike quality of life and narration at the Hall is a contrast to their one of travel and capitalist exertion.[59] The pantisocratic scheme advanced by Coleridge and Southey in 1794 for communal life in Pennsylvania also suggested that collective labor would leave many hours free for study. Thomas Poole writes of this setting that "a good library of books is to be collected and their leisure hours to be spent in study, liberal discussions, and the education of their children."[60]

Political Justice keeps company with these projects. Once our desire for luxury goods becomes less, Godwin argues, the trades supplying those goods will become redundant: "The object in the present state of society is to multiply labour, in another state it will be to simplify it" (432). The physical work necessary to supply all members of society with food and shelter is of a scale to be easily shared. Doing one's small part of this basic agricultural work is imagined as being to the benefit of everyone. Who, asks Godwin, "that sees the incessant industry exerted in this city and this island, and would believe that, with half an hour's industry *per diem*, we should not be every way happier and better than we are at present?" (433). The real aim of this arrangement, however, is that most hours of the day will be free of work, opening up in temporal terms a "clear and tranquil field in which every man shall be at liberty to discover and vindicate his opinion" (462).

Godwin imagines the time freed up by this new distribution of labor will be spent, at least in part, in studying books. He assumes that increased access to leisure will be an incentive to most to continue with education and intellectual innovation: "Leisure will be multiplied, and the leisure of a cultivated understanding is the precise period in which great designs, designs the tendency of which is to secure applause and esteem, are conceived. In tranquil leisure it is impossible for any but the sublimest mind to exist without the passion for distinction. This passion, no longer permitted to lose itself in indirect channels and useless workings, will seek the noblest course, and perpetually fructify the seeds of public good" (434). *Political Justice* does not specify exactly what people might read under these conditions: typically, Godwin's interest is in the tenor more than the content of future reading. But Godwin's emphasis is clearly on an ideal state of input and reflection rather than on literary output or new composition. He is not participating in the tendencies that Clifford Siskin's *The Work of Writing* ascribes to the same period, and which saw writing shored up as a form of professional, disciplinary, and masculine activity.[61] Godwin perceives that the writing of new books would only keep up the pressure to read them, whereas a utopian future is one in which such cycles of desire and demand will be broken. Thomas Northmore's *Memoirs of Planetes, or a Sketch of the Laws and Manners of Makar* (1795), a short utopian text inspired by *Political Justice*, solves this dilemma by having the residents read only short books, condensed statements of common truths: "Plain and Simple Facts, and energetic reasoning are their predominant features. I am sure my countrymen will be overjoyed to hear that some of their best works are comprised in one or two octavo volumes. Nay, I have frequently seen a small duodecimo that would put Hoadly to the blush."[62] For similar reasons, Godwin seems to imagine that the books being read in the future will be ones written and tested in the past. This means one will read them without fear of an author's irrelevance or of diverting hours from other kinds of work. Books will also become accessible enough in this context that all can pick them up in principle and put them down without a sense of urgency; they will be made to last, but not as specialized objects of cult devotion.[63]

Godwin's vision of reading in the future actually describes quite closely his own reading in the early 1790s. In his utopia, one will read old books in hours and quarters sequestered from the rest of the day and the household, in time set off from the time of work and of conversation. Although there is to be no private property, no bolts or locks, in his future state, there is to be privacy in which to study: "My apartment," he argues, "would be as sacred to a certain extent, as it is at present. No man would obtrude himself upon me to interrupt the course of my studies and meditation" (450). It is partly on these grounds that Godwin is opposed to cohabitation, certainly to large spaces of collective dwelling, but also, as we've seen, to versions of married life that would impinge on individual sanctuary. Social interaction will continue to be on one's own terms—at a point when one chooses to open one's door to the world, when one has outgrown adolescence, in the afternoon. Like Talbot's idyllic grotto, Godwin's study contains basic tools that might in principle be shared rather than precious or private papers; it is an environment anyone might enter and become a reader. Both Talbot and Godwin picture settings featuring a table, and what Godwin also describes as a "minimum of equipment." Ironically, of course, while Godwin imagines this as an extension of his reality, Talbot must figure her reading nook as a fantasy time-space different from any she ever occupies. Godwin's fantasy only involves his own setting and habits of time use becoming ones everyone might access in the future. Reading as he did, under conditions of leisure and serenity set off against the other hours of his productivity and social circulation, Godwin is propelled every morning into the future he is writing about.

Godwin's linking of this future to his own use of time and space in the present lines up with the way present and future are linked by the physicality of unread books. The books Godwin read in the mornings were being consumed in their authors' futures, much as the books he himself was writing would be consumed in his future. *Political Justice*, for instance, would play the role of the Restoration dramas Godwin was reading in the 1790s by being read serenely once people had won the time it claims for them. This dynamic takes us back to Stiegler, for whom writing allows us to weigh up the difference between now and what is to come in a way we couldn't do empty-handedly. The old book does not just record time in these terms: it intervenes in it as an object that seems to be ahead of us by just that distance measurable as a flicking of the fingers across pages one will have read one day. If, for Godwin, the books around him relate as fully to the future as they do to the past, it is because they serve in these terms as technical emissaries from a time to come, objects that are already in sight at the horizon where the future is happening. His claim that "the studious and reflecting only can be expected to see deeply into future events" is in this sense almost haptic (115).

Thus Godwin relies on the codex book as a tool in making sense of the way in which his own study exists in two times at once. There is the future, in which anyone may enter it as he does, and the present, in which he fights for their right to do so. The way the book splits between the time of its physical existence and the time of its reading, which is still to come, involves a similar chronotope. At the end of *Political Justice*, Godwin approaches this understanding when he describes "presages" being necessary to progress: "Mind cannot arrive at any great and illustrious attainment, however much the nature of mind may carry us towards it, without feeling some presages of its approach; and it is reasonable to believe that, the earlier these presages are introduced, and the more distinct they are made, the more auspicious will be the event" (470–71). In many of the scenarios he describes, books are these presages of the future. They are the objects and avatars of Muñoz's argument, oriented in time toward a future in which they already participate. Handling them involves recognizing them as models on a small scale of scenarios whose existence is embedded within *and* anterior to our own. But it also involves the realization that just by spending time with such objects, one is performing the arrival of that future one hopes for.

I have a colleague who reads in his office every day. The badly insulated walls of our 1960s building do not deter him; he has padded them with editions of Latin books and Renaissance texts. Even when he is on leave he sits there working peacefully while the rest of us dash between meetings and jammed copiers and office hours. From the crack in his door, which he keeps very slightly and affably ajar, the smell of leather wafts, and the details of a new edition of Latin letters he's been funded to write seem to emanate. At first I thought of him as a holdover from another era, an Oxbridge resident misplaced in the assertively forward-looking building in which he's ended up. But perhaps he's the real man of the future. When the revolution comes and the lights go off, his bags will already be packed: he knows what he'll be taking and what he'll be doing when we arrive. He has, in this sense, already set up camp in the place we're all frantically running around to get to.

HARDCOVER TRUTHS

Godwin did not write just books. In the 1780s, the bulk of his writing was for journals like the *New Annual Register* and the *Political Herald*. His pamphlet, *Cursory Strictures on the Charge Delivered by Lord Chief Justice Eyre to the Grand Jury* (1794), defending members of his circle against the charge of high treason, was as powerful as any of his novels and written in a form he commanded. In 1797, Godwin published *The Enquirer*, a series of essays prefaced by a reflection on the disadvantages of long systematic inquiry and acknowledging that "incessant recurrence to experiment and actual observation" might do just as well as the kind of systematic inquiry employed in longer works.[64] Early on in *Political Justice*, Godwin had already expressed doubt about whether books could really work as instruments of political change. Not only are they neglected by the majority of the population, he argues, "books to those by whom they are read have a sort of constitutional coldness. We review the arguments of an 'insolent innovator' with sullenness, and are unwilling to stretch our minds to take in all their force" (118). Godwin goes on to explain that conversation is a much more efficient medium of changing peoples' minds, one bringing "freedom and elasticity" to our mental disquisitions (118). His mood here foreshadows the one in which he writes and introduces his *Enquirer* essays as springing from conversation rather than textual engagement (vii). As David O'Shaughnessy shows, Godwin was also convinced for similar reasons of the value of theater as a better venue than books for changing opinions.[65] Drama, he claims in a note, provides "the link between the literary class of mankind and the uninstructed, the bridge by which the latter may pass over into the domains of the former."[66]

Yet, despite Godwin's sense of books' relative disadvantage as propaganda, *Political Justice* lists their relative advantages on other fronts: "Books have by their very nature but a limited operation; though, on account of their permanence, their methodological disquisition, and their easiness of access, they are entitled to the foremost place" (118). This helps explain why *Political Justice* was not simply a book, but a book published in two well-revised and relatively expensive quarto volumes. In a world of pamphlet literature, William St Clair notes, *Political Justice* stood out as "an honest attempt to take a longer view."[67] It was not, Godwin made clear when his admirer John Thelwell tried to lecture on its content, meant to be easily redacted.[68] The ripples it made were slowed down by its cost, with reading clubs and libraries accessing it more slowly than cheaper pamphlets that addressed similar topics, and with few workers who admired it being in a position to own it privately. For Pamela Clemit, the format in which *Political Justice* first appeared is evidence of Godwin's address to the "educated middling and higher classes of society, and confirmed its distance from the 'dangerous portability' of the occasional pamphlet."[69] And yet Godwin's account of books also brings his commitment to a democratic future into play in a way Clemit does not register here. True, books are not immediately compelling in the way news or conversation or good theater can be. As Godwin points out in *Political Justice*, the writer of books will have to wait for her influence to take effect: "In forming the mind of a young person, in endeavouring to give new bent to that person of mature years, I shall for a long time seem to have produced little effect" (123). But there is an advantage to books that comes from their being read slowly, not at the time of their being published, bought, or shelved, but that comes in Godwin's terms with the possibility, indeed the desirability, of their being read later. It is on these grounds, O'Shaughnessy writes, that "Godwin's model must be considered temporally—his vision was of a gradually widening participation in the production of truth as society generates more leisure time for reading and reflection. Thus, Godwin may have been exclusionary, but he was always anxious that rational-critical debate should move toward and envelop everyone."[70]

The duration that comes with books of classic status, whose relevance can be both still to come and already there, is in this sense a generous and inclusive one. It can be connected, as it is in Ian Balfour's argument throughout *The Rhetoric of Romantic Prophecy*, to early nineteenth-century ways of approaching the Bible as a book oriented in descriptive terms, but also in anticipation of its audience, toward things still to come. Even an expensive book can be seen in this light as more democratically accessible than the pamphlet or sermon: relieved of the urgency of needing to be read now, it awaits the moment at which all will have time and space to access it. The reading of a book as something that must inevitably lag behind its composition becomes in this case its merit, an incentive to read it in the "unplugged" mode that Godwin favored. In contrast, a pamphlet, as Godwin professed, might actively incite interest in the present without offering enlightenment in the long run.[71]

But it is also worth noting that Godwin was quick to dismiss hasty novel reading on these same grounds. His own first novel, *Things as They Are; or, The Adventures of Caleb Williams* (1794), aimed to take advantage of fiction's appeal. Godwin uses it to lay out his political argument against tyranny and misrepresentation in the form of a case study where Caleb challenges his former employer

Falkland to reveal the truth of a murder he has committed before the time of the novel begins. As novelist, Godwin plays with the already-thereness of Falkland's crime as a fact we learn of early on by putting the narrative emphasis on the ways the secret does and doesn't percolate through an unjust society. Like the book, the narrative is already there: it needs only to be read. Reading in pursuit of the narrative thus upsets, in Godwin's terms, a balance that he wanted to hold in favor of the book's consumption as a key element of the truth it has to tell. This becomes clear when he reports on his reaction to hearing how one reader has enthusiastically consumed it:

> And, when I had done all, what had I done? Written a book to amuse boys and girls in their vacant hours, a story to be hastily gobbled up by them, swallowed in a pusillanimous and unanimated mood, without chewing or digestion. (Joseph Gerrald) told me that he had received my book late one evening, and had read through the three volumes before he closed his eyes. Thus, what had cost me twelve month's labour, ceaseless heartaches and industry, now sinking into despair, and now raised and sustained in unusual energy, he went over in a few hours, shut the book, laid himself on his pillow, slept and was refreshed and cried, Tomorrow to fresh pastures new.[72]

Godwin's despair at the idea that *Caleb Williams* could be read in a single sitting speaks, Philp comments, "to the extent to which he was ambitious for his work, not as entertainment, but as instruction and communication."[73] But it also suggests just how clearly his understanding of the difference between these categories depended in his terms on instructive reading as something that needs substantial time devoted to it, rather than on books as devices allowing one to forget or speed up time.

The other advantage Godwin hopes, perhaps overoptimistically, to win by making *Caleb Williams* and *Political Justice* books is the space to lay out a comprehensive and complete case for political justice. As the preface to the *Enquirer* registers, the method of argument pursued by the author of a book may be rife with disadvantages, but it is "the highest style of man," appropriate to the observation of "immense and distant objects" (vi). When *Political Justice* states that it is "the duty of individuals to publish truth without diffidence or reserve, to publish it in its genuine form without seeking aid from the meretricious arts of publications," Godwin has books and their slow and repeat reading in mind (462). Telling truth whole, he goes on, strengthens its effect: "The more it is told, the more it is known in its true dimensions, and not in parts, the less it is possible that it should coalesce with or leave room for the pernicious effects of error" (462). Godwin is not directly criticizing other media platforms here, and yet he worries that an agglomeration of opinion pieces and pamphlets cannot have the same effect as a book in laying out the truth in a single, coherent, and reusable form. A book that tells the truth whole may not be read as true. It may not be read at all, at least not until people are at leisure to engage it. As Northmore reassures himself at the end of *Memoirs of Planetes*: "Sow the seeds of justice and truth among your countrymen. Sow them deep. And the noisome weeds that now overspread the land and seem to choke their growth, will be totally lost in that strength and energy with which they will in their due time vegetate into maturity."[74] The book remains for Godwin a privileged player in the long game of telling the truth, holder over time of the foremost place in the hierarchy of media.

The concrete advantages of a book's ability to lay out a whole truth before people are ready to hear it endear the form to Godwin, but other writers of the period looked to different media to secure a future readership. Wordsworth, unhappy with the critical reception of his later poetry, appeals to a not yet existent but rightful reader of his poems. Turning aside with disdain from "this multitude of unhappy and misguided, and misguiding beings" who only "dip into books in order to give an opinion of them," he imagines a work of regeneration that "must be a work of time."[75] Many of his late poems for this audience are staged as written on stone, not paper, suggesting a material investment in posterity that anticipates a period of dormancy for the text rather than its gradual dissemination. As stone, the text simply waits for the moment in the future when it will be discovered.[76] This same idea fuels "The Wreck of Westminster Abbey," a 1788 satire bearing the false publication date of 2001 and presented from the perspective of a twenty-first-century reader discovering the gravestones of "the most conspicuous personages who flourished towards the latter end of the eighteenth century."[77] Barbauld's "Eighteen Hundred and Eleven" (1812), which delivers up a vision of a Britain reduced by war to the ruins of a fallen empire, has a similar eye to a future reader with whom the mistakes being made now are to be discovered on the surface of the earth, "by time's slow finger written in the dust."[78] There is no sense in any of these cases that a gradual clearing of the way for reading and understanding of texts must have happened in order for the poem's truth to be recognized in the future. Although the break from the present anticipated by Barbauld is in some ways a slow one, bringing the long arc of history into play, for the reader of the future, epiphany happens suddenly and in connection with the poem as a relic of the past.[79]

Godwin's investment in future reading also differs from the spirit in which writers around the turn of the century invoke the future as a source of uncertainty. In *Modernity's Mists: British Romanticism and the Poetics of Anticipation* (2015), Emily Rohrbach argues for Keats, Shelley, and Austen as among those harnessing literature to the future. The materiality of the book is implicated here, as in Derrida's and Currie's discussions of the future anterior, and as it was in the last chapter, as a literal point of contact with contingency. "The concept and poetics of anticipation," Rohrbach contends, "are decidedly nonteleological. Rather than suggesting a linear movement toward a specified end point or goal, the mist of anticipation opens the present up to multiple possibilities."[80] Rohrbach is contending with Koselleck's case for all teleological narratives from the eighteenth century becoming routed in new ways toward an unknowable future. She argues more specifically for the less standard modes of narration, coupled to the contingencies of literature's existence, opening up forms of uncertainty more interesting than explicit statements of doubt. For instance, the paper text's amenability to being reread and its openness to being approached anew become in her reading a case for revisiting what might more conventionally be understood as Austen's nostalgia for a time past. Rather than suggesting that a book captures the past or prefigures the future, Rohrbach sees it as implanting itself in time with a preparedness to hazard the long climate of uncertainty.[81]

But Godwin, as we have seen, is quite certain about the form the future should take; he is confident, even, of its resembling in important ways the life that takes shape each morning in his study. Books in his terms are not instruments of contingency. They are presages of the point at which readers will arrive once they have time to read them. As they measure the difference between now and that future, books anticipate not so much a huge shift in perspective or historical events but the course of a struggle that must happen if everyone is to have a chance to read *Political Justice* (or Naomi Klein). How far away is the future measured in this way? Several mornings sealed off in one's study from daily life? A sabbatical? A lifetime? And how does the calculation work if everyone is to be included? The book is nothing, of course—a mere 400 pages standing between now and the future. But as a political project, the unread book anticipates the longest *durée*, the most ambitious of all political horizons. The point at which we will all have found the time to read the books we want to read (including the ones we may not yet know we want to read): this is the point at which Godwin is aiming, and he's clearly right to perceive it as a utopian goal.

As a PhD student, I spent a year doing research in Berlin. The wall had just come down and the city was full of things that were out of place. There was a jumble of old objects to be found out of their historical order, and a feeling of history being tumbled around with them. Thanks to generous exchange rates, we had enough money for beer, but not really enough for books. Nevertheless, my German partner went out one day and came back with the complete works of

Marx and Engels in their classic "Blaue Bände" edition. He'd found all forty-three volumes for sale at a market. How could one resist? We used them to build a coffee table, and they were reason enough to build shelves. Now they have been around the world with us several times. I keep paying to ship them even though I still read German only slowly and find time for original sources to be increasingly scarce. Sometimes we have used them to look things up—though in one move or another, the volume containing the index got damaged (the insurance people were at a loss when it came to working out what to pay us). But still: a whole edition. One day I will read them through. When that has happened, the future will be here.

YOU CAN'T SKIP PAGES

If, as A *Simple Story* and Godwin's view of the book both suggest, a good happy ending (to a novel, to a life, to history) can be in view from the outset, why is reading needed at all? The possibility of eliminating time from reading is on the table in twenty-first-century discussions of technologies that might allow us to read faster, search better, or outsource reading altogether. It was also raised in the eighteenth century, as we've seen at a number of points in the last chapters—with Richard Griffith's idea, for instance, that reading a text partially or non-sequentially might release time back to the reader in the form of contingency, or in my resistance to Soni's emphasis on trial narratives requiring us to follow characters' fates through once, from beginning to end. Developments in book history including abridgments, indexes, and anthologies suggest that the desire of the reader to economize on the time it takes to get a text into one's head has been longstanding. Schopenhauer already remarked long ago that buying books would be better if you could also buy the time to read them.[82] And yet, as I approach the close of my own book, I want to promote Godwin's emphasis on the necessity of the time it takes to read a book, and on the years it might take for a book to find its audience.

Godwin's position that justice is something that must be gradually endorsed rather than revealed (as, say, news, or feeling, or poetry engraved upon stone) is related in complicated ways to the change he hopes a reader will undergo by being given time with a book. For Godwin, real change, as his despair over a fast reading of *Caleb Williams* suggests, takes time. *Political Justice* argues explicitly that "one of the principal means of information is time." The time involved in conveying a fact becomes just as important here as that involved in conveying a story. "Our Time" Godwin argues in this same passage, is the "theatre" in which our faculties and possessions can "unfold themselves": "There is nothing the right disposal of which is more sacred" (298). As he stresses the time information needs to reach its public—to be read—Godwin advances an eighteenth-century world view that goes well beyond the concerns with media I've focused on here, but which nevertheless engages the physical logic of the book.

Ian Balfour has argued for the performative logic that both Godwin and Inchbald invoke when they oppose in principle the form of the contract—the social contract and the marriage contract. Both authors, he suggests, see the commitment to a future state as wrong when it forecloses the process of discovery involved in getting there. This relates to the case made in *Political Justice* against marriage as "the worst of all laws," an institution that teaches us in its contractual form to shut our eyes to inquiry rather than to learn from experience (446). If, as I argued earlier, *Amelia Opie* subtly takes up this case by promoting rereading as a good thing, *Political Justice* and *A Simple Story* do so in Balfour's account much more explicitly, and on more political grounds, by advocating the difference that time makes. For instance, in the seventeen years that pass between *A Simple Story's* two parts, everything changes: Miss Milner / Lady Elmwood has an affair with Lawnly, the man she'd rejected for Dorriforth, and Dorriforth finds this out and ceases to love her. The five years of Dorriforth's absence are not narrated in the novel, but they register radically as the unspeakable interregnum in which legitimate changes of heart occur. Reading and education can be seen conservatively as the things that might have prevented Miss Milner's rash behavior, but in reality when they are given time in the novel, it's as the time in which contracts and promises are unbound. Inchbald and Godwin can be seen to share on these grounds, and in Balfour's argument, an antipathy to anything that would commit us to conclusions not properly unfolded in time. "The temporality of knowledge" argues Balfour, "need not converge with the arc of futurity cast by the promise, and in Godwin's economy, knowledge must take priority over the speech-act, even if the promise comes first in time."[83] Both Godwin and Inchbald advocate the recognition of time as something revolutionary when it is given to education, and to books, because it entails a commitment to the future that is different from the present. If spending time with books involves bringing a new world into being it is not by rearranging spatial coordinates (which are already in view), but by changing the way we use time. This is not a contract, as Balfour stresses, but the performative nature of the book's promise.

This helps underscore the circular logic at play in Godwin's larger argument for leisure time being both the outcome and the prerequisite of real revolution. Embracing this circularity, *Political Justice* claims of the opportunity for "leisure and reflection" that "if it did not precede the general dissemination of truth would at least be the immediate result of it" (22). Godwin forgoes here any straightforward sense of causality in favor of a much more complex account of how change will occur. For the best and most rational arguments around us to come into effect, people must have the opportunity to read and absorb them at leisure. But in order for this to occur, the democratic future represented in those arguments must already have arrived. This makes another kind of time—a leisurely time, an untimely time, a queer time, a time that is not counted in terms of industry or profit—both the thing Godwin is urging us to advance toward, and the thing the books around us now require us to participate in. The performative logic is similar to the one Muñoz deploys in arguing for what it would mean to inhabit a queer future: "To live inside straight time and to ask for, desire, and imagine another time and place is to represent and perform a desire that is both utopian and queer. To participate in such an endeavour is not to imagine an isolated future for the individual but instead to participate in a hermeneutic that wishes to describe a collective future, a notion of futurity that functions as a historical materialist critique."[84] The future here is one it is necessary to "perform" rather than just "imagine." The difference between the future as a fantasy and as a reality, constituted for both Muñoz and Godwin as a difference in the way time is experienced, closes up as it is read. Leisure time, the means by which more things would be read by more people, moves into view as soon as reading succeeds: if this message is circular, is also laden with hope.

For many months during 2016 and 2017, I stopped reading books. The pounding rhythm of world news became all-consuming. Every day a new scandal kept me scrolling away through the pages of the papers and the websites and pressing on the links. It's just as Lackington said, but worse. But I did not stop buying books. Just now, in fact, I have bundles of them newly purchased in sight, keeping me company as the promise of what I want to do next. And I know my students make the same piles—texts gathered, if not as books, then certainly as the files they store away like squirrels in the cloud for a summer day. This is a dynamic that those who think real revolutions must happen in the rapid fire of communication, churned up by action, cannot easily account for. Applauding our digital turn, Serres suggests that text searching has freed up time from reading the classics, clearing the way for the real revolution that must happen in the here and now.[85] But he forgets that the material form of the book—the bound and printed book meant to last, the book conspicuously already there, the book whose doneness is its very form but whose reading is always still to come—was never commensurate, either with the act of looking backward, or with the operation of reading as one that focuses on the present. Pages and volumes offer us images of hours we hope one day to have spent, words we hope one day to have read. It is for this reason, A *Simple*

Story and *Political Justice* both suggest, that the medium of the book supports narratives oriented toward a world still to come, one where there will be time to read, later, what we are busy writing and doing now. This is a model of space and time that the materiality of the codex book and the temporality of its reading performs with a particular, material tenacity. But it is also a way of imagining the future that can be taken up more generally, in any of the less papery forms in which we might still allow ourselves to imagine stockpiling text. Book reading has been seen by many in media theoretical terms as the retracing of the tracks that old writers laid down; book history has been seen by still more as an understanding of the way lives and texts unfolded in the past. But the boundness of books has never been just a sign that they are over: it has always also been a sign of that time for reading that is still to come.

Academic Time

One way reading was approached in the eighteenth century was through the university, which could be seen as giving expert readers the time that others lacked. Samuel Johnson describes the role of the academic in these terms: "An academic is a man supported at the public cost, and dignified with public honours, that he may attain and impart wisdom. He is maintained by the public, that he may study at leisure; he is dignified with honours, that he may teach with weight. The great duty of an academic is diligence of inquiry, and liberality of communication. Of him that is appointed to teach, the first business is to learn, an unintermitted attendance to reading must qualify him to be heard with profit."[1] The solution wasn't perfect, of course. Most were excluded from this life of reading, and many might have questioned—as they continue to question—the "liberality of [the] communication" that flowed back from the universities to the public on the reading being done there. Chad Wellmon has argued persuasively that cultures of disciplinary research stemmed anxieties around the too-muchness of books in this period, but for most people eighteenth-century universities did not help directly with the too-littleness of time any more than twenty-first century universities seem to do today.

In the longer view, however, there is a history of universities and adult educational institutions becoming places that would give ordinary people more time to read, places where whole years of students' lives could be legitimately devoted to books. Much of this book was written in Denmark, a country where all university students are still paid by the government to study for five years. UK and US universities have never harbored readers quite so safely, but in the 1960s and 1970s, unprecedented levels of state funding helped open tertiary education to many at or below the class positions of the readers I've discussed in this book. Carter and Turner might, for instance, have found their places there as bookish students. At this time, humanities degrees were central to the project of university education, and "reading" for a degree really did involve handling books in the way Talbot imagined as ideal. In the reality of state-funded and democratic education, the desires of many of the eighteenth-century readers I've described here found their shape as a social project. A future different from the one Godwin foresaw but in many ways compatible with his hopes *did* arrive.

While writing this book I have sometimes suspected myself of nostalgia, something that gets attached easily these days to recollections of reading. In fact, I am not nostalgic about books as physical objects. I give my books away freely and always have. I am promiscuous with editions, and fairly indifferent to the special features of books as objects. I do not particularly want to smell or touch them. Homing in on their materiality in this project has been a way to think about the particular set of practices associated with consuming them in time, and situating ourselves in time through our relation to them, rather than a way of attributing special powers to bound papers. Imagining people as readers has allowed me to think about their efforts to situate themselves in time more generally, and to appreciate the creativity of their efforts as pitched against those that were applied at the same time in history to regulate, measure, and monetize time. It has also been a way to give historical coordinates to a certain set of behaviors associated with books in particular, and to disassociate those from the other kinds of reading supported by print culture and our ubiquitous engagement with screens today. But the fact that the reading of codex books can and almost certainly will pass away does not move me particularly. All the practices I have described here in relation to print reading—the partitioning of time, the returning to a text over a lifetime, the contact with a range of possibilities, and the anticipation of a quieter future—are ones that might continue through the reading of digital or online texts. I hope we can build machines that measure up to books in their ability to run at different speeds, promote intervals, and provide deep kinds of access as well as shallow. We have yet to invent the machines that turn themselves off when darkness comes, or allow one to access the same page in different ways, or send us to a quieter place on Sunday, or anticipate a better future—but there's no reason to think we can't.

So my nostalgia is not really directed toward paper books. But I do hope for a politics that would see humanist education in its widest sense as the reason for wanting and making time. One of the things that became obvious in researching this project was how many eighteenth-century people saw reading in these terms, as the thing they would spend more time doing if they had the choice. This is the view that Talbot takes of evenings in which she is forced to socialize rather than to be left alone in the study, and it's the view Turner takes when he envies the wealthy, not for their money, but for their time to study. It is what draws Carter to Epictetus, and Grenville to Aristotle, both of them on the lookout for theorists that allow them to justify reading, like happiness, as its own end. And it is what Inchbald and Talbot sense as the disadvantage facing women, who, unable to claim time alone with a book, feel deprived of the chance to arrive at the real end point of a good life. By the logic of any of these arguments, you don't read in order to become more independent, or to move up the social ladder: you pine for this independence and movement in order to be able to read. The one person I have mentioned who puts this otherwise is Lackington, for whom education is something he promotes as a way of making a place for himself in the world. But as I suggested at the very beginning of this study, the real facts of Lackington's biography show that, much like Grenville and Inchbald, even his reading was, in reality, the reward for and not the means to wealth.

A good society, a post-revolutionary one in the Jacobin terms of the late eighteenth century, has figured in this study in terms of the creation and equitable distribution of that time. And for centuries, on the larger scale, this is the kind of time that seemed to be just on the horizon. It is there as Wordsworth writes *The Prelude*, anticipating a better and quieter world to come, and it is there in 1967 as E. P. Thompson celebrates *The Prelude*, protecting the university as an institution that might remain into the future a place in which work time-discipline does not apply. If it is no longer easy to imagine humanities education as the thing that might be an alternative to work and an incentive to think of the end of work, then something has been lost (which is not to say that there are not other books to be written about reading as work, and about the ways in which thinking instrumentally about the time it takes has been part of an important working class project).

To put this in explicitly political and particularly British terms, the stories of eighteenth-century readers presented here as stories of creative time use and life projects designed to keep in view reading as its own reward provide an interestingly stark contrast to the idea of education based on reading as a means to an end, an investment of time that would pay for itself when other goals are met. The modern view of the university, the one that has grown since the 1970s, suggests more grotesquely than ever that the time spent studying the humanities is instrumental to moving a society toward greater levels of profit. In 1970, just a few years after writing "Time, Work-Discipline, and Industrial Capitalism," E. P. Thompson edited *Warwick University, Ltd.*, a publication responding supportively to recent student unrest. In his own contribution to that book, Thompson directly questions the university's mantra that "by far the most important educational products of a university must be those which go to reinforce a system which in fact is directed by criteria of profitability."[2] Along with such a view, he suggests, comes the general tendency to imagine time spent with books as instrumental to other kinds of individual and social success.

The inverse logic, which Stefan Collini tries out in *What Are Universities For?*, states that "rather than saying that extending human understanding is valuable because it provides the means to prosperity we should surely say that one of the reasons prosperity is valuable is because it provides the wherewithal to extend human understanding."[3] This is the reasoning that's applied by almost all the readers I've looked at in their giving time to and making time for reading. As objects that conspicuously need time, books in any form make sense only if given time, so the possibility and promise of their reading becomes an argument for the reorganization of the way we think about the monetization of time, work, the long life, and the future. As something that will happen when time is made and given, reading becomes here the crowning achievement, not just of the good life, but of a society that gets things right for the majority of its citizens.

Systems theory does not, of course, have much to say about such political goals or institutional programs, but it does help bring into focus the way in which it is people and objects being fed through time that gives any system its character. Take, for example, Collini's point about the way in which disciplines like history, once introduced within universities as vocational and useful, began to seem less useful, or more than useful, once students read for them. "It is not," he argues, "the subject-matter itself that determines whether something is, at a particular moment, classed as 'useful' or 'useless.' Almost any subject can fall under either description. Rather, it is a question of whether enquiry into that subject is being undertaken under the sign of limitlessness."[4] Luhmann might put this differently. He might say that the difference between a system designed to move you somewhere and one that allows you to enjoy having arrived boils down to the different ways in which the spending of time is viewed within that system. Latour might say that the difference between a mode in which texts are deemed useful and one where they are lauded as useless lies in how they are handled: as things that must make something else happen, or as events in their own right. Godwin might say it is the tendency of the reader, and the distance he can take from his own moment, that matters in one case and the products of education in the other. But all of these accounts resonate with what I've been arguing, which is that book reading should be understood in time and as a quality of time use. Herein lies the stumbling block for those who study the history of reading in empirical terms, as a reaction of a person to a particular text. But it's also a block for readers today who struggle with the idea of books as unjustifiably time consuming. If reading is merely *useful*, the less time it takes the better. But from another perspective, it is duration that helps constitute reading's allure as a utopian activity and its promise as a pleasurable one.

Put simply, and in contemporary terms, the portable lesson might be this: if we want our kids and our students to keep reading books (loaded as that wish is, in all the ways we might want to examine), it's not enough to give them better texts or faster ways to extract and home in on information. We must also encourage them to value reading as an activity that is explicitly time consuming and productive in multiple ways, some of which may not be visible in the short term or through the lens of economic gain. That is the precondition of reading books as I've explored it here. It is also its highest aim—the one it sets for civilization as having a form beyond work that might actually arrive.

It is customary to thank organizations that have funded the writing of a book in the acknowledgments. But the organizations that have sponsored this book are woven too closely into the fabric of its argument not to mention here. I taught through the years I wrote it in institutions that were generous with leave, under the protection of tenure-track or tenured positions. The University of Michigan, Copenhagen, and Warwick (despite Thompson's justified critique) are all institutions that have given me time in generous measure. In the first year I was working on this project, I held an Alexander von Humboldt fellowship and in the year I finished it, a Leverhulme Trust grant. The years away from teaching and administration gifted by these grants facilitated absolutely the reading on which *Reading and the Making of Time in the Eighteenth Century* is based. I am no Lackington: without those years, my own reading would not have happened.

In this incarnation, I may be part of the last generation of humanist academics to hold grants that come as terms away from other kinds of work, rather than as funding for events, networking, and business trips. The money that trickles our way these days, especially in Europe, tends to do so in the form of cluster funding and project grants, initiatives that keep people busier than ever—collaborating, writing, scanning, and data mining—but initiatives that are much less easily associated with reading. Perhaps there is a need in our new world for this kind of collaboration between humanists, but what I have written is a history of scholars and readers wanting and making and hoping for time to read, and so I am conscious of my good luck in having had that time. Whatever the substratum of texts in the future, let us keep reading them. And whether we continue to read books, let us remember the aim that their shape helped render palpable: that of making societies that give more people more time to be readers.

Notes

INTRODUCTION: When Do We Read?

1. Two important manifestos associated with this claim are Nicholas Carr's *The Shallows: What the Internet Is Doing to Our Brains*, and Jonathan Crary's *24/7: Late Capitalism and the Ends of Sleep*. As a counterpoint to these, see Natalie Phillips, *Distraction: Problems of Attention in Eighteenth-Century Literature*, which argues, in a spirit kindred to the one in which I discuss time use, for attending to texts having been a problem long before digital media, and for the idea that attention was once deeper than it is now as a back-formation of our own moment.

2. For a good account of the endurance of books, see Striphas, *The Late Age of Print*.

3. Harvey, *The Condition of Postmodernity*, 239.

4. Muñoz, *Cruising Utopia*.

5. Wajcman, *Pressed for Time*.

6. Hansen, "Living (with) Technical Time." Hansen argues, for instance, that "digital inscription yields a time that is not constituted but *given*, a time that *gives itself* for myriad and potentially incompossible temporalities" (302).

7. I am quoting here from Koselleck, *The Practice of Conceptual History*, 165. For the larger form of this argument, see Koselleck's *Futures Past*.

8. Foucault, *Discipline and Punish*.

9. Thompson, *English Working Class*, 451.

10. Thompson, "Time, Work-Discipline, and Industrial Capitalism," 90. Hereafter cited parenthetically.

11. I am drawing here on Jeff Knight's unpublished paper, "English Literature and the Imagined Library," in which he claims, "Book historians and textual materialists at present seem almost reflexively to write from the premise that an unprecedented, destabilizing increase in book production took place in the period they study." The central text here is Blair's *Too Much to Know*, but similar claims form the premise of Piper's *Dreaming in Books*.

12. See Raven, *The Business of Books*, and Raven, *Bookscape*.

13. Take, for instance, Raven's reporting of the multiplying of newspaper sales, which grew from 7.3 million copies sold in 1750 to 16 million sold in 1790. In 1760, Raven reports in the *Business of Books*, London has four daily papers and five or six that publish three times a week, but by 1790, there are thirteen morning, one evening, seven three-a-week, and two twice-a-week (257–58).

14. Among the most influential studies of reading in this period is St Clair's *Reading Nation,* which draws directly on historical rather than literary evidence. Although many of St Clair's conclusions about the formations in which reading happened and the importance, say, of reading across a lifetime and reading out of order of publication are resonant with what I argue, St Clair draws largely on publishers' records, a source that many have found problematic. His method has been contested, for instance by Fergus, whose *Provincial Readers in Eighteenth-Century* is much more meticulously focused on records of what people actually borrowed and purchased. Mee's *Conversable Worlds* is a good example of a study that deploys both kinds of sources, documenting different zones of contact with books and relying on a range of sources.

15. Rose, *Intellectual Life.*

16. Altick, *English Common Reader,* 85.

17. Anderson, *Imagined Communities,* 35.

18. For the most famous description of the spaces of eighteenth-century London, see Habermas, *The Structural Transformation,* 1–56. Habermas continues to emphasize spaces (over, say, the availability of leisure time to workers or the temporal rhythms of publication) in more recent pieces such as "Equal Treatment of Cultures," 1–28.

19. See Warner, *Publics and Counterpublics,* 175–77.

20. Lynch, *Loving Literature,* 154.

21. Sherman, *Telling Time,* 5.

22. Pasenek and Wellmon, "The Enlightenment Index." See also Multigraph Collective, *Interacting with Print.*

23. See, for instance, Price, "Search," 93–97.

24. Wellmon, *Organizing Enlightenment.*

25. *Low-Life,* 8. Hereafter cited parenthetically.

26. Pratt, *Archives of American Time,* 44.

27. Hoggart, *The Uses of Literacy*; Radway, *Reading the Romance.*

28. Steedman offers a full account of the publication history of this piece in "Cries Unheard, Sights Unseen," 67.

29. de Certeau, *The Practice of Everyday Life,* 174.

30. Garrett, "Subterranean Gratification," 120.

31. To highlight just some of these studies: In *Provincial Readers,* Fergus focuses on catalogs of provincial booksellers. Bannet's *Transatlantic Stories* highlights the actual reception of stories documenting transatlantic life in terms of their reprinting and circulation on both sides of the Atlantic. In *The Child Reader,* Grenby not only describes in detail the diverse material children had access to but also sheds some light on the rituals of their bedtime and Sunday reading. In "'Books Without Which I Cannot Write,'" Susan Staves highlight the importance of women's borrowing and exchanging books. Sher's *Enlightenment and the Book* brings the nexus of cooperation between authors and publishers into focus. Keymer and Sabor write of commercial uptakes and transformations of Richardson's work in *Pamela in the Marketplace.*

32. For projects of this kind, see *Mapping the Republic of Letters* (http://republicofletters.stanford.edu); *Literary Atlas of Europe* (http://www.literaturatlas.eu/en/); *The Grub Street Project* (http://grubstreetproject.net); and *Six Degrees of Francis Bacon* (http://www.sixdegreesoffrancisbacon.com).

33. Derrida, *Given Time,* 41.

34. In *Outline of a Theory of Practice*, Bourdieu argues at the anthropological level that observing a pattern of behavior as if it were objective and reversible, one must forget that "it is all a question of style, which means in this case timing and choice of occasion for the same act—giving, giving in return, offering one's services, paying a visit, etc.—can have completely different meanings at different times" (6).

35. Derrida, *Given Time*, 41.

36. Serres, with Latour, *Conversations*, 109. Hereafter cited parenthetically.

37. Serres, *The Birth of Physics*.

38. Serres, with Latour, *Conversations*, 109.

39. Luhmann, *Art as a Social System*, 21. Hereafter cited parenthetically.

40. For the best overview of Luhmann's theory, see Luhmann, *Social Systems*.

41. *The Guardian*, April 18, 2007.

42. Luhmann, *Soziologische Aufklarung*, 38. My translation.

43. Luhmann, *Love as Passion*.

44. Latour, *Inquiry into Modes of Existence*, 224. Hereafter cited parenthetically.

45. Cited in de Certeau, *Practice of Everyday Life*, 69.

46. Warner, "Uncritical Reading," 29–30; Brewer, "Reconstructing the Reader."

47. For a description of Johnson reading when he should have been at church, see Boswell, *Life of Johnson*, 30. I discuss Godwin's time habits at length in chapter 4, where I draw on his diary. *The Life of Thomas Holcroft* includes the diary alongside several descriptions of Holcroft's childhood reading (89). For Thrale's time shortage, see Sherman's description in *Telling Time*, 251–52.

48. Cobbett, *Advice to Young Men*, 35–36.

49. This is described by Montagu Pennington, Carter's nephew, in his *Memoirs*, 22.

50. Gibbon, *Memoirs of My Life*, 72.

51. Griffith, *Genuine Letters*, 2.397.

52. The Borrowers' Register at the Innerpeffray Library has been transcribed by Katherine Halsey and her team. She reports on this material in "A Quaint Corner of the Reading Nation."

53. Halsey, Looser, Warner, and Lynch have all worked on the quirky and fantastic reception of Austen's fiction. Tadmor's essay on Thomas Turner's reading appears as "'In the Even My Wife Read to Me.'"

54. See Derrida, *Writing and Difference*.

55. Good local meditations on the changing contexts for reading today include McGill and Parker, "The Future of the Literary Past," and Piper, "Reading's Refrain."

56. Kittler, *Gramophone, Film, Typewriter*.

57. Stiegler, "Memory," 79.

58. For Knight's description of Lackington's bookshop, see Raven, *Bookscape*, 137. For more detail on Lackington see Bankes, "James Lackington," 157–74; Landon, "Small Profits Do Great Things," 387–99; and Altick, *English Common Reader*, chapter 2.

59. Lackington, *Memoirs*, 420–21. Hereafter cited parenthetically. Lackington's claim is vigorously refuted by Altick (*English Common Reader*, 39–40) and by Raven, who warns against book historians taking such autobiographical claims too seriously (see "New Reading Histories," 268–87).

60. Raven, "Selling One's Life,"1–23.

61. For review of catalog numbers and discussion of the rates at which Lackington inflated these see Bankes, "James Lackington," 162.

62. Bankes also discusses some of these notes in her dissertation, remarking that they constitute an unusual use of catalog space (171–74).

63. Lackington, *Lackington's Catalogue for 1797*.

64. Lackington, *Confessions*, 24–25.

65. Muñoz, 25.

CHAPTER 1: Time Divided

1. Crary, 24/7, 8–9.

2. Crary, 74.

3. See, for instance, Freeman, "Synchronic/Anachronic," 129–43.

4. Carter, *Letters*, 1:353. Hereafter cited parenthetically.

5. Turner, *Diary*. Hereafter cited parenthetically.

6. Knox, *Winter Evenings*. Hereafter cited parenthetically.

7. Osborne, *The Politics of Time*, 17.

8. For an anthropological approach to multiple temporalities, see Greenhouse, *A Moment's Notice*. A recent example of a study focusing on "domestic time" as one cyclical event within modernity is Damkjær, *Time, Domesticity and Print Culture*.

9. Henkin, "Hebdomadal Form," 52.

10. Pratt, *Archives of American Time*, introduction.

11. Nightingale, *Cassandra*.

12. Carter and Talbot, *A Series of Letters*, 2.250.

13. Thrift, "Owners' time and own time."

14. Schellenberg provides a strong account of Talbot's role in Elizabeth Montagu's coterie in *Literary Coteries*, 84–90.

15. See Zuk's introduction to *Bluestocking Feminism*, 4 and 9.

16. See Major, "Life and Works of Catherine Talbot"; also Myers, *Bluestocking Circle*, 64–90. For the argument that Talbot's concerns are more elegiac, see Rasmussen, "'Speaking on the Edge of My Tomb': The Epistolary Life and Death of Catherine Talbot."

17. Clarke, "Bluestocking Fictions," 460–78.

18. Talbot, *Journal* [27]. Hereafter cited parenthetically.

19. Talbot, *Reflections*, 66. Hereafter cited parenthetically.

20. Carter writes here to Eliza Berkeley in 1787. The unpublished letter is held at the British Library among Berkeley's correspondence (BL Add. MS 39312) [53].

21. Gibbon, *Memoirs of My Life*, 94.

22. Talbot's "Dialogue 1" is reprinted in Zuk's *Bluestocking Feminism*, and I am quoting here from this edition, 109.

23. Talbot's "Essay" is reprinted in Zuk's *Bluestocking Feminism*, and I am quoting here from this edition, 72.

24. Durkheim, *Elementary Forms of Religious Life*, 328; Leach "Two Essays," 124–36.

25. Zerubavel, *The Seven Day Circle*, 120.

26. Henkin, "Hebdomadal Form," 63.

27. Browne, *Sunday Thoughts*, 116.

28. Perovic, *The Calendar in Revolutionary France*, 58.

29. Zerubavel makes this point in *The Seven Day Circle*, where he describes the invention of the week as "one of the first major attempts by humans to break away from being prisoners of nature and create an artificial world of their own" (4).

30. See Cunningham, *Leisure in the Industrial Revolution*, 42. For evidence of the state of the Anglican Church in the mid-eighteenth century, see, for instance, Bowen, *Yorick's Congregation*.

31. Research drawing on court testimony to show the long hours that Londoners worked in the 1700s confirms that working hours in the city increased by one fifth during the second half of the century; the research attributes this increase largely to the loss of Monday as a holiday. See Voth, "Time Use," 497–99.

32. McCrossen, *Holy Day, Holiday*, 16.

33. Kilner, *A Course of Lectures*; Bellamy, *The Family Preacher*, 1.xi.

34. For an example of a text written for Sunday school, see *Hints for the Institution*. For an appeal to stricter laws, see *Sabbath Keeping*.

35. Talbot's "Sunday" was published in the *Rambler*, no. 30, Saturday, June 30, 1750. It is reprinted in Zuk's *Bluestocking Feminism*, and I am quoting here from this edition, 41.

36. Kilner, *Letters from a Mother*, Vol. 2, 72.

37. Talbot's approach to the issue is Christian, but it resonates, for instance, with Immanuel Kant's 1780 advocacy of enlightenment as something that happens after work, once one's hours of obedience are over for the day. The whole idea of the public sphere is in this sense premised on one's day having different temporal modes.

38. Barbauld, *Defense of Public*, 31. This was first published in 1792 as "Remarks on Gilbert Wakefield's Inquiry into the Expediency and Propriety of Public or Social Worship." For more on the controversy see Mandell, "Prayer, Feeling, Action," 117–42.

39. *The Grounds of Complaint*.

40. Hamersley, *Advice to Sunday Barbers*, 23.

41. Coke, *Letters and Journals*.

42. Wakefield, *Enquiry*.

43. See, for instance, Priestley's *Letters to a Young Man*. The "pleasing and useful distinction in our time" is defended before arguments being made for the civilization of mankind and the suggestion that individually decided days of worship would have a weaker effect. None of these arguments are directly religious (48).

44. See Lynch's account of Johnson in *Loving Literature*, where she describes his complex but also fastidious relationship with reading poetry, 44–50.

45. Boswell, *Life of Johnson*, 624. Hereafter cited parenthetically.

46. Quoted in Miller, *The Peculiar Life of Sundays*, 103.

47. Boswell and Temple, *Correspondence*, 378.

48. Temple, *Diaries*, 23. Hereafter cited parenthetically.

49. Bettany, "Memoir of W. J. Temple," xlii.

50. Raven, "Arrangements for Reading," 175–201.

51. Temple, like Turner, is worth comparing to the diarists that Henkin describes in "Hebdomadal Form" as relying on the seven-day cycle to structure their reflections and provide the unit of time in which the past seems to have happened.

52. Turner, *Diary*. This edition includes an Appendix D, "Thomas Turner's Reading."

53. Tadmor has written on the Turners' reading practices, confirming that Turner really reads only during the day on Sundays, but noting that there are some exceptions when his business is quiet. See "'In the even,'" 162–74.

54. See Schellenberg's description in *Literary Coteries*, chapter 1.

55. Richardson, *Grandison*, iii.xviii. Hereafter quoted parenthetically by volume and letter.

56. Price, "The Executor's Hand," 331–42, 337.

57. Jones, *Consensual Fictions*, 69.

58. See, for instance, Grandison organizing space to read and answer letters while excusing himself from supper (iii.xviii).

59. Latour, "On Interobjectivity," 228–44.

60. Serres, *Hermes: Literature, Science, Philosophy*, 75.

61. Sherman, *Telling Time*, 143.

62. Dames, "The Chapter: A History."

63. Derrida, *Given Time*, 41.

64. See, for instance, Talbot's letter to Carter on April 4, 1754, extolling the virtues of Harriet from *Grandison* (2:166).

65. See, for instance, Schwarz, "Addicted to Distraction."

CHAPTER 2: Joining Up Time

1. Small, *Value of Humanities*, 120.

2. Lynch, *Loving Literature*, 149.

3. Lynch, 152.

4. See, for instance, "Review" in *The Monthly Review*, 18. All references to *Epictetus* are to this first edition.

5. Carter's translations into English prior to *Epictetus* consisted of *An Examination of Mr. Pope's Essay on Man* from the French of M. Crousaz and *Sir Isaac Newton's Philosophy Explain'd for the Use of the Ladies*, from the Italian of Francesco Algarotti. In 1738 she authorized publication of her slim collection, *Poems on Particular Occasions*. Some of these poems then appeared without her permission, notably in Robert Dodsley's *Collection of Poems by Several Hands*. See Hawley, "Carter, Elizabeth (1717–1806)."

6. See, for instance, Eger's *Bluestockings*, which is generally emphatic about the merits of mid-century feminist writing. Eger builds on the work of recovery done by Myers (*Bluestocking Circle*) and Hawley in her edition of Carter's works (*Bluestocking Feminism*). Bigold's *Women of Letters* argues not only for Carter's investment in her writing but in her control of her posthumous reputation.

7. Montagu to Carter, November 2, 1767. Quoted in Eger, *Bluestockings*, 94.

8. Pennington, *Memoirs*, 22. Hereafter cited parenthetically.

9. Quoted in Pennington, 136. This same letter describes the rope leading from a bell at her bedroom window to the ground floor of her house, on which a sexton was instructed to pull between four and five each morning, and which Virginia Woolf refers to in *A Room of One's Own*.

10. Warner, "Uncritical Reading," 23.

11. Scott, *Literary Translation*, 80.

12. Benjamin, "The Task of the Translator," 263.

13. In *Women and Enlightenment,* O'Brien summarizes Carter, not as celebrating Epictetus's principles as correct, but as situating him as setting "an example to modern readers of the enormous capacity and the limits of human reason in the moral sphere" (62). On the other hand, O'Brien goes on, Carter finds Epictetus's life "an admirable example of the divinely implanted instinct for moral striving" (62).

14. Talbot writes to Carter that she has heard a rumor of her being offered this place as tutor to the princess's children in April 1754 (2:166).

15. Reports that this was Carter's view in her old age are offered in Bigold, *Women of Letters,* 198. Carter was also singled out in Hayley's *Old Maids,* where she is mentioned as a particularly dignified example of old age and invoked as protectress of the project (see Looser, *Women Writers,* 164).

16. See O'Brien, *Women and Enlightenment.* Commenting on the hostile tones of Carter's footnotes, O'Brien suggests that "readers did and still do wonder what drew her to Epictetus in the first place" (62).

17. Benjamin, "The Task of the Translator," 257.

18. Soni, *Mourning Happiness.*

19. Soni, 209.

20. Soni, 239.

21. This is a loss with which literary critics have been concerned in other registers. Segal writes beautifully in *Out of Time* of the need for a different perspective on the lifetime, as does Small in *The Long Life.*

22. Letter to Miss Highmore, April 28, 1752, in Carter, *Elizabeth Carter,* 143.

23. See Talbot's letter to Carter on March 16, 1751, in which she assumes that the two share a high estimation of *Clarissa* (2:16).

24. Recent work by Lynch has suggested that reading books straight through might not have been the norm in the Georgian period, where she locates the culture of scrapbooks and commonplace books.

25. Jupp, *Lord Grenville,* 14.

26. For an account of Grenville's activity in this period, see Jupp, 294.

27. These notebooks are held in the British Library manuscript collection, where they are cataloged among the Dropmore papers (Grenville, "Notes on Reading of Winter," "Notes on Aristotle," and "Reading of Plutarch").

28. Clergymen were often advised in the period to takes notes on what they had read by summarizing its content in their own words. This exercise was seen to improve their style as well as retention of the text's content.

29. Referring to his journal of 1803, Jupp reports on Grenville's "wide and intent" reading for that first year of his actual retirement having included "the last ten books of the Odyssey (read twice); the poetry of nearly a score of other classical poets; the Greek version of the New Testament (also read twice) and the first five chapters of the Epistle to the Romans; a selection of English sermons and biblical commentaries including Richard Bentley's Epistles of Pharlaris; and finally, a number of French and English history books concerned mainly with the seventeenth century" (298).

30. Grenville, "Reading of Plutarch" [16].

31. Grenville, "Notes on Aristotle" [1].

32. See Soni, *Mourning Happiness,* chapter 1.

33. I am quoting here from John Gillies's translation of Aristotle, which Grenville

was reading hot off the press (*Ethics and Politics*, 1.ii). All subsequent translations and parenthetical citations are also from this edition of the *Ethics*.

34. Grenville, "Notes on Aristotle" [25–26].

35. For a detailed historical account of the role of men's bookkeeping in the period more generally, see Harvey, *The Little Republic*. It is also interesting to think about the way in which Grenville's use of the page to record his reading relates to the practice of double-entry bookkeeping, which Mary Poovey (*Modern Fact*) situates as central to modern subjectivity.

36. Small, *Value of Humanities*, 122.

37. Opie, *Adeline Mowbray*, 41. Hereafter cited parenthetically.

38. Eger, *Bluestockings*, 169–72.

39. Godwin, *Memoirs*,105.

40. See Butler, *Jane Austen*, 121, and Kelly, "Discharging Debts," 198–203.

41. See, for instance, King, "Portrait of a Marriage," 27–62, especially p. 41 for evidence of Opie's reading Godwin.

42. Two articles that take this more moderate position include Eberle, "Amelia Opie's 'Adeline Mowbray,' "121–52, and Howard, " 'The Story of the Pineapple,' " 355–76.

43. Balfour, "Promises, Promises," 225–50.

44. Godwin, *An Enquiry Concerning Political Justice*, 446. Hereafter cited parenthetically.

45. See, for instance, Sheridan, "Being Last," 173–86.

46. St Clair, *Reading Nation*, 4.

47. Anne McWhir's introduction to Opie's *Adeline Mowbray* also discusses the textual history of the novel more generally, and the differences in the two editions (33).

48. She is writing here to Mrs. Duncome, London, May 29, 1801 (Carter, *Unpublished Letters*, 193).

49. Looser, *Women Writers*, 20. See also Small, *The Long Life*.

CHAPTER 3: Other Times

1. A *Series of Genuine Letters between Henry and Frances* was published in three installments. Volumes 1 and 2 were published in 1757, Volumes 3 and 4 in 1766, and Volumes 5 and 6 in 1770. Unless otherwise stated, parenthetical citations in the rest of the chapter are to the second edition (2.397).

2. Quoted in Pasanek, *Metaphors of Mind*, 10.

3. Pasanek, 17.

4. Pegge, "The Novel's Proper Use," 240.

5. Panagia, *Impressions of Hume*, 11.

6. Kermode, A *Sense of an Ending*, 7, 45–46.

7. Kermode, 133–36.

8. Krämer, "Cultural Techniques," 93–109, 97.

9. McGann, *Radiant Textuality*. See also Birkerts, *Gutenberg Elegies*.

10. Quoted in Ernst, *Digital Memory and the Archive*, 30. See also Manovich, *The Language of New Media*. Manovich represents the changes in data storage as part of a seismic shift in the way we approach narrative generally, one that moves us away from text as narrative to seeing all narrative as data. See also Harpold, "Contingencies."

11. For a polemical and rather untechnical version of this argument, see Serres, *Thumbelina*. Serres sees the very format of the page and the book having been associated with spatial regulation and institutional hierarchy: "The space of the lecture hall was designed as a field of forces whose orchestral center of gravity was the stage, with its focal point at the lectern, which was literally a *power point*" (34).

12. Piper, *Book Was There*.

13. See Chartier, "Languages, Books, and Reading," 133–52; see also Stallybrass, "Books and Scrolls," 42–45.

14. A good point of contrast here is the audiobook, a form that is currently growing in popularity thanks to Audible, but which has a long history of being frustrating to move around in and which is much harder to manipulate in this respect than the paper book. See Rubery, *Untold History*.

15. Pasanek and Wellmon, "The Enlightenment Index."

16. Peters, *The Marvelous Clouds*, 290.

17. Stoiceff and Taylor, "Introduction," 3. For more recent work drawing attention to the page as a unit of paper see Mak, *How the Page Matters*, and Gitelman, *Paper Knowledge*.

18. Brown, *The Pilgrim and the Bee*, 65.

19. Levine, *Forms*, 41.

20. Wellbery, "Contingency," 249.

21. Price, *How to Do Things*.

22. Sterne, *Tristram Shandy*, 252.

23. See Molesworth, *Chance*. For Molesworth's reading of the chance scene in *Tristram Shandy*, see 193.

24. Kidgell, *The Card*, 1:13.

25. Shebbeare, *The Marriage Act*, 1:33.

26. Fielding, *Tom Jones*, 543.

27. Colman, *Polly Honeycombe*, 4.

28. Haywood, *Miss Betsy Thoughtless*, 1.11.

29. See, for example, Warner's description in "Uncritical Reading" of Mary Rowlandson's using her Bible as a book to be opened randomly at a page of advice.

30. For a good treatment of the progressive/digressive movement in *Tristram Shandy*, see Drury, "The Novel and the Machine," 337–42.

31. Harvey, "Manuscript History," 1–21.

32. See Gitelman, *Paper Knowledge*, 29–35.

33. Austen, *Northanger Abbey*, 185.

34. Austen, *Emma*, 339.

35. Austen, *Mansfield Park*, 73.

36. Fleming, "'Kannitverstan,'" 72–78, 79.

37. This is the overarching argument of Perry's *Novel Relations*.

38. I am thinking here of Perry but also Walker's *Marriage, Writing, and Romanticism*.

39. O'Connell, "Vicars and Squires," 385.

40. Cleland, "Review of 'Amelia,'" 510–12.

41. Lennox, *The Female Quixote*, 327.

42. Fielding, *Amelia*, 15. Hereafter cited parenthetically.

43. Molesworth, *Chance*, 176.

44. Zomchick, *Family and the Law*, 131, and Castle, *Masquerade and Civilization*, 242.

45. Potkay, "Determination and Agency," 335–58.

46. Mark McGurl, "Being and Time-Management," talk given at the English Institute, September 2016.

47. Sheridan, *Sidney Bidulph*.

48. Griffith, *Posthumous Work*, 1.173. This text was later appended to real editions of Sterne's work (see Sidney Lee, "Griffith, Richard").

49. Griffith, *Something New*, 2:2. Hereafter cited parenthetically.

50. There is much more biographical information on Elizabeth than Richard. See Eger, "Griffith, Elizabeth (1727–1793)," and Rizzo, "'Depressa Resurgam,'" as well as Ricciardi and Staves, "Introduction."

51. Griffith, *Young Married Women*, 27.

52. Griffith, *The Delicate Distress*, 4.

53. In this, Griffith responds to *Amelia*. See Ballaster, "Contexts, Intertexts, Metatexts," 347–58.

54. Panagia, *Impressions of Hume*, 6, 63.

55. Panagia, 11.

56. To see this argument taken up most fully, see Meillassoux, *After Finitude*, 65.

57. Rasch, *Niklas Luhmann's Modernity*, 58.

CHAPTER 4: Time to Come

1. Posnock, "I'm Not There," 85–95, 86.

2. Peters, *The Marvelous Clouds*, 317. See also Taleb's argument for the "antilibrary," which he depicts as a collection of unread books productively out of proportion with the books one has read (*The Black Swan*, 1).

3. This is cited as the epigraph to André Gorz, *Paths to Paradise*.

4. See, for instance, Weeks, *The Problem with Work;* Haug, "The Four-in-One-Perspective," 119–23; and Livingston, *No More Work*. These texts advance very different arguments for the twenty-first-century reduction in paid work but share the assertion that this is the direction we must take in the future.

5. Grahame, *The Sabbath*, x.

6. Muñoz, *Cruising Utopia*, 22–23.

7. Muñoz, 26.

8. Zuk, *Bluestocking Feminism*, "Dialogue 1," 107.

9. See McKenzie, *Bibliography*.

10. McGann, "Philology in a New Key," 327–46, 334.

11. Ernst, *Digital Memory*, 122.

12. Stiegler, *Technics and Time*, 172.

13. See Derrida and Prenowitz, "Archive Fever," 9–63, 18. For a reading of the materiality of the book that is oriented toward Heidegger's future, see Lurz, *Death of the Book*, which argues, very differently from Stiegler, that novels partake materially in time as an orientation toward their own death.

14. Stiegler, *Technics and Time*, 231.

15. Luhmann, "The Future Cannot Begin," 130–52.

16. Works emphasizing the role of media in this period include Siskin and Warner's edited collection *This is Enlightenment*, which is organized around this very point; Favret's *War at a Distance*, which focuses on the role of mail and news; Levy's *Family Authorship*, on the circulation of unprinted manuscript; Warner's *Protocols of Liberty*, which combines Latour and media systems analysis; and Burkett's *Romantic Mediations*, which shows how romantic media forms anticipate later ones to follow.

17. See Barrell and Mee, "Introduction."

18. Knox, *Winter Evenings*, 1: C3.

19. Beresford, *The Miseries of Human Life*, 172. I am grateful to Deidre Lynch for this reference. For an argument about the way in which binding transformed pamphlets into books in the eighteenth century, see Knight, *Bound to Read*.

20. Lackington, *Memoirs*, 419.

21. "The Adventures of a Robinson Crusoe," 185–92, 185.

22. Currie, *The Unexpected*, 12.

23. Currie, *About Time*, 15.

24. Currie, 150.

25. See, for instance, the way Radway reports in *Reading the Romance* (199) on readers who flick casually to the end of a book to see if they approve of the ending before reading it.

26. I use the future anterior as a tense advisedly. It's important not only for Currie's argument in *About Time*, but for Derrida's on the future orientation of the book in *Paper Machine*, where he suggests, without any particular emphasis on the passage of time involved in reading, that paper supports this idea of a future in which the present will have become the past.

27. See Jenkins, *I'll Tell You What*, 10.

28. Inchbald, *Memoirs of Mrs. Inchbald*, 1:50.

29. The claim that she doesn't own books comes from Boaden, editor of *Memoirs of Mrs. Inchbald*, 132. He says she borrowed them from friends. This is corroborated by her diaries, in which such loans are mentioned.

30. This is Boaden's summation of the day (see *Memoirs of Mrs. Inchbald*, 83).

31. Inchbald, *Diaries*,1:260.

32. Inchbald, 1:280.

33. For details of this period see Boaden's piece in Inchbald, *Memoirs of Mrs. Inchbald*, 271–74; Jenkins, *I'll Tell You What*, 289–98; and the Introduction to Robertson, *Elizabeth Inchbald's Reputation*.

34. Inchbald to Godwin, Nov 24, 1792. Quoted in Jenkins, *I'll Tell You What*, 317.

35. In *Masquerade and Civilization*, Castle calls *A Simple Story* the best novel written in the period, one whose full meanings critics are still unpacking.

36. See Jenkins, *I'll Tell You What*, 273.

37. Inchbald, *A Simple Story*, 29–31. Hereafter cited parenthetically.

38. Kelly suggests that we lose interest in the first part of the novel despite the neatness of the construction (*The English Jacobin Novel*, 92). For those invested in the revolutionary tenor of the novel, the second half has generally been seen as disappointing. See, for instance, Rodgers, "Inhibitions," 63–78. However, at least two more recent readings offer more positive accounts of the relationship between the two parts. Min ("Giving Promises," 105–27) takes direct aim at the idea that the second part is not up

to par by suggesting that it's only with the second part that omens of the first are realized. And Frank has recently offered another way to think about joining them up, as two stories of male reform that come into focus if we see Dorriforth/Lord Elmwood as Inchbald's key character ("Melodrama," 707–30).

39. Anderson, "Revising Theatrical Conventions," 5–30, 25. For other commentary on the role of the theater, see Frank, who argues in her forthcoming book on melodrama for a complex hybrid of melodrama and the novel, in which Inchbald emerges as ultimately committed to the politics of the novel.

40. Bakhtin, "Forms of Time," 255. We might compare this to Genette's claim that "the narrative text, like every other text, has no other temporality than what it borrows, metonymically, from its own reading" (*Narrative Discourse*, 34).

41. Spencer, "Introduction" to Inchbald, *A Simple Story*, xx.

42. Frank, "Melodrama," 707–30.

43. None of Inchbald's biographers (Boaden, Jenkins, Robertson) have represented her as genuinely revolutionary, all tending to see her as a fairly passive witness to male energies. It's been left to literary critics Gary Kelly and Terry Castle to assert just how many rules she breaks in her fiction.

44. Batchelor's *Women's Work* describes carefully the stakes women had in the definition of writing as a form of work in this period. I do not want to suggest that Inchbald's later writing could not be figured in these terms as a form of labor, or that she didn't want it to be seen as such; only that her doing it during a period of her life in which her financial and domestic security were fairly settled took the pressure off her having to define it in these terms.

45. Godwin, *An Enquiry Concerning Political Justice*, 22. Hereafter cited parenthetically.

46. See Meyers, O'Shaughnessy, and Philp, "Introduction."

47. From Godwin, *Diary*.

48. See Meyers, O'Shaughnessy, and Philp, "Introduction."

49. Quoted in St Clair, *The Godwins and the Shelleys*, 189.

50. See Lamb's financial plea on his behalf in *Letters*, letter 286.

51. See Godwin, *Diary*, "Reading."

52. Godwin reports on visiting Lackington's in his diary but only in the years after Lackington had handed it over to his cousin (See Godwin, *Diary*, July 7, 1813, and June 10, 1813). He socialized with the new proprietor, George Lackington.

53. Godwin, *Letter of Advice*, 11.

54. Godwin, 9.

55. Godwin, *The Enquirer*, 129.

56. Philp describes Godwin's extreme patterns of sociability in "Unconventional Calling."

57. Mee's "The Press" presents Godwin as fairly socially reserved. O'Shaughnessy, on the other hand, emphasizes Godwin's moving toward theater and the genre of the essay, and his openness by the turn of the century toward new kinds of publics, a turn of which his novel writing becomes evidence ("*Caleb Williams*," 423–48).

58. Godwin, *The Enquirer*, 149.

59. However, see also Batchelor's description in chapter 2 of *Women's Work* of how important work in this community was.

60. Poole's summary is quoted in Adelman, *Idleness, Contemplation and the Aesthetic*, 103. In the French tradition, see Louis-Sébastien Mercier's *Memoirs of the Year Two Thousand Five Hundred*, in which the books that have survived into the future combine abridged versions (of Voltaire, Montesquieu), full versions (of Rousseau) and the *Encycopedie*, which serves schoolchildren everything they need to know (chapter 28). In the nineteenth century, William Morris's *News From Nowhere* also addresses the role of reading in a utopian setting. His visitor to the future learns that printing presses are winding down, so that it is old books, circulating in no particular rush, to which children find their way. In settings such as the central cloister, people are seen reading in a posture of relaxation.

61. Siskin, *The Work of Writing*.

62. Northmore, *Memoirs of Planetes*, 132.

63. As advocates of reading, Godwin and Northmore could be contrasted here with the other literary connoisseurs of the period, the antiquarians, book collectors, and literary fans that Lynch identifies in *Loving Literature* as the first "lovers" of literature as a specialized object.

64. Godwin, *The Enquirer*, xi.

65. O'Shaughnessy discusses this specifically in *William Godwin and the Theatre*.

66. Godwin, *Letters*, 66.

67. St Clair, *The Godwins and the Shelleys*, 87. See also Altick's description in *The English Common Reader* of Paine's circulation (70).

68. For an account of this relationship and Godwin's critical response to Thelwall's lectures see Allen, "William Godwin's Influence," 662–82.

69. Clemit, "Godwin, Political Justice," 86–100, 86.

70. O'Shaughnessy, "*Caleb Williams*," 423–48.

71. Philp, "Preaching to the Unconverted," 1–18, 10.

72. Quoted in Philp, 5–6.

73. Philp, "Preaching to the Unconverted," 6.

74. Northmore, *Memoirs of Planetes*, 140.

75. Wordsworth, "Letter to Lady Beaumont."

76. On this topic, see Simonsen, *Wordsworth and Word-Preserving Arts*, and Bennett, *Romantic Poets*.

77. Croft, *The Wreck of Westminster Abbey*.

78. Barbauld, *Selected Poetry and Prose*, 214.

79. For a reading of the poem that situates it in media terms, see Sachs, "The Glimmer of Futurity," 17–30.

80. Rohrbach, *Modernity's Mists*, 3.

81. Rohrbach's argument is in agreement here with Sachs's for the late-eighteenth-century interest in classical forms as revealing a deeper acknowledgment of the fact that the future is actually unlikely to resemble the past ("Scales of Time," 697–718).

82. Peters, *The Marvelous Clouds*, 321.

83. Balfour, "Promises, Promises," 225–50, 229. Here Min's "Giving Promises" includes a useful response to Balfour. A promise, she argues, is not the same as a contract because it is not binding in the same way, and can be fulfilled only at its point of delivery. This, Min argues, entices Inchbald to use the promise rather than the contract as a way of ventilating *A Simple Story's* characters to uncertainty even while locking

them into cross-generational relationships they do not entirely control. In these terms, the book operates as a promise rather than a contract: it secures its end point in advance of all parties agreeing to ride its course; it leaves open the question of whether it will be read.

84. Muñoz, *Cruising Utopia*, 26.

85. Serres, *Thumbelina*.

CODA: Academic Time

1. "On the Character and Duty of an Academick," a two-page text attributed to Samuel Johnson, was published as the appendix to an obscure work by John Moir, *Hospitality*.

2. Thompson, *Warwick University Ltd.*, 162.

3. Collini, *What Are Universities For?*, 111.

4. Collini, 55.

Works Cited

Adelman, Richard. *Idleness, Contemplation and the Aesthetic, 1750–1830*. Cambridge: Cambridge University Press, 2011.

"The Adventures of a Robinson Crusoe. Written by Itself." *The Young Gentleman's and Lady's Magazine, or Universal Repository of Knowledge, Instruction and Amusement* 1 (1799): 185–92.

Allen, B. S. "William Godwin's Influence on John Thelwall." *PMLA* 37.4 (1922): 662–82.

Altick, Richard. *The English Common Reader: A Social History of the Mass Reading Public, 1800–1900*. Chicago: University of Chicago Press, 1957.

Anderson, Benedict. *Imagined Communities: Reflections on the Origin and Spread of Nationalism* [1983], rev. ed. London: Verso, 1991.

Anderson, Emily Hodgson. "Revising Theatrical Conventions in 'A Simple Story': Elizabeth Inchbald's Ambiguous Performance." *Journal for Early Modern Cultural Studies* 6.1 (2006): 5–30.

Aristotle, *Ethics and Politics*. Translated by John Gillies. London: A. Strahan, T. Cadell, and W. Davies, 1797.

Austen, Jane. *Emma* [1815]. Edited by James Kinsley. Oxford: Oxford University Press, 2008.

———. *Mansfield Park* [1814]. Edited by James Kinsley. Oxford: Oxford University Press, 2008.

———. *Northanger Abbey* [1817]. Edited by James Kinsley. Oxford: Oxford University Press, 2008.

Bakhtin, Mikhail. "Forms of Time and of the Chronotope in the Novel." In *The Dialogic Imagination: Four Essays*. Translated by Caryl Emerson and Michael Holquist. Austin: University of Texas Press, 1981.

Balfour, Ian. "Promises, Promises: Social and Other Contracts in the English Jacobins (Godwin/Inchbald)." In *New Romanticisms*. Edited by David Clark and Donald Goellnicht. Toronto: University of Toronto Press, 1994: 225–50.

———. *The Rhetoric of Romantic Prophecy*. Palo Alto: Stanford University Press, 2002.

Ballaster, Ros. "Contexts, Intertexts, Metatexts: Eighteenth-Century Prose by Women." *Eighteenth-Century Fiction* 11.3 (1999): 347–58.

Bankes, Sophie. "James Lackington (1746–1815) and Reading in the Late Eighteenth Century." Unpublished Dissertation: Open University, 2013.

———. "James Lackington (1746–1815): Reading and Personal Development." In *The History of Reading*. Vol. 2, *Evidence from the British Isles, c.1750–1950*. Edited by Katie Halsey and W. R. Owens. London: Palgrave Macmillan, 2011: 157–74.

Bannet, Eve Tavor. *Transatlantic Stories and the History of Reading 1720–1810: Migrant Fictions*. Cambridge: Cambridge University Press, 2011.

Barbauld, Anna Letitia. *Defense of Public or Social Worship by the Late Mrs. Barbauld*. Bristol, UK: William Browne, 1829.

———. *Selected Poetry and Prose*. Edited by William McCarthy and Elizabeth Kraft. Peterborough, ON: Broadview, 2001.

Barrell, John, and Jon Mee. "Introduction." In *Trials for Treason and Sedition, 1792–1794*. 8 Vols. Edited by John Barrell and Jon Mee. London: Pickering and Chatto, 2006.

Batchelor, Jennie. *Women's Work: Labour, Gender, Authorship*. Manchester: Manchester University Press, 2010.

Bellamy, Daniel. *The Family Preacher: Consisting of Practical Discourses for Every Sunday Throughout the Year*. 2 Vols. London: B. Law, 1776.

Benjamin, Walter. "The Task of the Translator." In *Walter Benjamin: Selected Writings*, Vol. 1. Edited by Marcus Bullock and Michael W. Jennings. Cambridge, MA: Belknap Press, 1996.

Bennett, Andrew. *Romantic Poets and the Culture of Posterity*. Cambridge: Cambridge University Press, 1985.

Beresford, James. *The Miseries of Human Life: or, The Groans of Samuel Sensitive, and Timothy Testy*. London: William Miller, 1806.

Bettany, Lewis. "Memoir of W. J. Temple." In *Diaries of William Johnston Temple, 1780–1796*. Edited by Lewis Bettany. Oxford: Clarendon, 1929.

Bigold, Melanie. *Women of Letters, Manuscript Circulation, and Print Afterlives in the Eighteenth Century*. Basingstoke, UK: Palgrave Macmillan, 2013.

Birkerts, Sven. *The Gutenberg Elegies: The Fate of Reading in an Electronic Age*. New York: Fawcett Columbine, 1994.

Blair, Ann. *Too Much to Know: Managing Scholarly Information before the Modern Age*. New Haven: Yale University Press, 2011.

Blake, William. *Milton a poem, and the final illuminated works*. Edited by Robert Essick and Joseph Viscomi. Princeton: Princeton University Press, 1993.

Blanchot, Maurice. *The Book to Come*. Translated by Charlotte Mandell. Palo Alto: Stanford University Press, 2002.

Boswell, James. *Life of Johnson* [1791]. Edited by R. W. Chapman. Oxford: Oxford University Press, 1970.

Boswell, James, and William Temple. *The Correspondence of James Boswell and William Johnson Temple, 1756–1795*. Edited by Thomas Crawford. Edinburgh: Edinburgh University Press, 1997.

Bourdieu, Pierre. *Outline of a Theory of Practice*. Translated by Richard Nice. Cambridge: Cambridge University Press, 1977.

Bowen, Martha. *Yorick's Congregation: The Church of England in the Time of Laurence Sterne*. Newark: University of Delaware Press, 2007.

Brewer, John. "Reconstructing the Reader: Prescriptions, Texts, and Strategies in Ann Larpent's Reading." In *The Practice and Representation of Reading in England*.

Edited by James Raven, Naomi Tadmor, and Helen Small. Cambridge: Cambridge University Press, 1996: 226–45.

Brown, Matthew. *The Pilgrim and the Bee: Reading Rituals and Book Culture in Early New England.* Philadelphia: University of Pennsylvania Press, 2007.

Browne, Rev. Mr. Moses. *Sunday Thoughts. In Three Parts. Adapted to the Various Parts of the Christian Sabbath: and its Different Duties, Exercises, and Employments.* London: A. Millar, 1753.

Burkett, Andrew. *Romantic Mediations: Media Theory and British Romanticism.* Albany, NY: SUNY Press, 2016.

Butler, Marilyn. *Jane Austen and the War of Ideas*. Oxford: Clarendon, 1975.

Carr, Nicholas. *The Shallows: What the Internet Is Doing to Our Brains*. London: Norton, 2011.

Carter, Elizabeth. *Bluestocking Feminism: Writings of the Bluestocking Circle, 1738–1785*, Vol. 2. Edited by Judith Hawley. London: Pickering and Chatto, 1999.

——. *Elizabeth Carter, 1717–1806: An Edition of Some Unpublished Letters*. Edited by Gwen Hampshire. Newark: University of Delaware Press, 2005.

——. *Poems on Particular Occasions*. London, 1738.

——. Unpublished letter to Eliza Berkeley, 1787. (BL Berkeley Papers. Add. MS 39312).

Carter, Elizabeth, and Catherine Talbot. *A Series of Letters between Mrs. Elizabeth Carter and Miss Catherine Talbot, from the Year 1741 to 1770*. Published from the Original Manuscripts in the Possession of Rev. Montagu Pennington. 4 Vols. London: J. Rivington, 1809.

Castle, Terry. *Masquerade and Civilization: The Carnivalesque in Eighteenth-Century Culture and Fiction*. Palo Alto: Stanford University Press, 1986.

Chapone, Hester, and Catherine Talbot. *Bluestocking Feminism: Writings of the Bluestocking Circle, 1738–1790*. Vol. 3. Edited by Rhoda Zuk. London: Pickering and Chatto, 1999.

Chartier, Roger. "Languages, Books, and Reading from the Printed Word to the Digital Text." *Critical Inquiry* 31.1 (2004): 133–52.

Clarke, Norma. "Bluestocking Fictions: Devotional Writings, Didactic Literature and the Imperative of Female Improvement." In *Women, Gender and Enlightenment*. Edited by Barbara Taylor and S. Knott. Houndsmill, UK: Palgrave Macmillan, 2005: 460–78.

Cleland, John. "Review of 'Amelia.'" *The Monthly Review*, December 1751: 510–12.

Clemit, Pamela. "Godwin, Political Justice." In *Cambridge Companion to the British Literature of the French Revolution in the 1790s*. Edited by Pamela Clemit. Cambridge: Cambridge University Press, 2011: 86–100.

Cobbett, William. *Advice to Young Men, and, Incidentally, to Young Women, in the Middle and Higher Ranks of Life*. London: B. Bensley, 1829.

Coke, Lady Campbell. *Letters and Journals of Mary Campbell Coke*. Edited by J. A. Holme. Edinburgh: David Douglas, 1889.

A Collection of Poems by Several Hands. Edited by Robert Dodsley. London: R. Dodsley, 1748.

Collini, Stefan. *What Are Universities For?* London: Penguin, 2012.

Colman, George. *Polly Honeycombe, a Dramatick Novel of One Act*. London: 1760.

Crary, Jonathan. *24/7: Late Capitalism and the Ends of Sleep*. London: Verso, 2014.
Croft, Sir Herbert. *The Wreck of Westminster Abbey, Being a Selection from the Monumental Records of the Most Conspicuous Personages Who Flourished towards the Latter End of the Eighteenth Century* [1788?]. London: 2001.
Cunningham, Hugh. *Leisure in the Industrial Revolution c.1780–c.1880*. London: Croom Helm, 1980.
Currie, Mark. *About Time: Narrative, Fiction, and the Philosophy of Time*. Edinburgh: Edinburgh University Press, 2007.
——. *The Unexpected: Narrative Temporality and the Philosophy of Surprise*. Oxford: Oxford University Press, 2013.
Dames, Nicholas. "The Chapter: A History." *The New Yorker*, October 29, 2014.
Damkjær, Maria. *Time, Domesticity and Print Culture in Nineteenth-Century Britain*. Houndsmill, UK: Palgrave Macmillan, 2016.
de Certeau, Michel. *The Practice of Everyday Life*. Translated by Steven F. Rendall. Berkeley: University of California Press, 1984.
Derrida, Jacques. *Given Time: 1. Counterfeit Money*. Translated by Peggy Kamuf. Chicago: University of Chicago Press, 1991.
——. *Paper Machine*. Translated by Rachel Bowlby. Palo Alto: Stanford University Press, 2005.
——. *Writing and Difference*. Translated by Alan Bass. London: Routledge, 1978.
Derrida, Jacques, and Eric Prenowitz. "Archive Fever: A Freudian Impression." *Diacritics*, 25.2 (1995): 9–63.
Drury, Joseph. "The Novel and the Machine." *The Eighteenth Century* 42.2 (2009): 337–42.
Durkheim, Émile. *Elementary Forms* of Religious Life. Oxford: Oxford University Press, 2001.
Eberle, Roxanne. "Amelia Opie's 'Adeline Mowbray': Diverting the Libertine Gaze; or, the Vindication of a Fallen Woman." *Studies in the Novel* 26:1 (1994): 121–52.
Eger, Elizabeth. *Bluestockings: Women of Reason from Enlightenment to Romanticism*. London: Palgrave Macmillan, 2010.
——. "Griffith, Elizabeth (1727–1793)." *Oxford Dictionary of National Biography*. Oxford University Press, 2004; online edition, May 2009.
Epictetus. *All the Works of Epictetus, Which Are Now Extant; Consisting of His Discourses, Preserved by Arrian, the Enchiridion, and Fragments*. Translated by Elizabeth Carter. London: Andrew Millar, John Rivington, and Robert and James Dodsley, 1758.
Ernst, Wolfgang. *Digital Memory and the Archive*. Minneapolis: University of Minnesota Press, 2013.
Favret, Mary. *War at a Distance: Romanticism and the Making of Modern Wartime*. Princeton: Princeton University Press, 2010.
Fergus, Jan. *Provincial Readers in Eighteenth-Century England*. Oxford: Oxford University Press, 2007.
Fielding, Henry. *Amelia* [1751]. Edited by Martin Battestin. Oxford: Clarendon, 1983.
——. *The History of Tom Jones, a Foundling* [1749]. Edited by Thomas Keymer. London: Penguin, 2005.

Fleming, Paul. "'Kannitverstan': The Contingent Understanding of Anecdotes." *Oxford German Studies* 40.1 (2011): 72–89.

Foucault, Michel. *Discipline and Punish: The Birth of the Prison*. Translated by Alan Sheridan. New York: Random House, 1975.

Frank, Marcie. "Melodrama and the Politics of Literary Form in Elizabeth Inchbald's Works." Eighteenth-Century Fiction 27.3–4 (2015): 707–30.

———. *The Novel and the Theatrical Repertory from the Restoration to Jane Austen*. Lewisburg, PA: Bucknell University Press, forthcoming.

Freeman, Elizabeth. "Synchronic/Anachronic." in *Time: A Vocabulary of the Present*. Edited by Joel Burges and Amy J. Elias. New York: New York University Press, 2016: 129–43.

Garrett, Matthew. "Subterranean Gratification: Reading after the Picaro." *Critical Inquiry* 42:1 (2015): 97–123.

Genette, Gerard. *Narrative Discourse: An Essay in Method*. Translated by Jane Lewin. Ithaca, NY: Cornell University Press, 1980.

Gibbon, Edward. *Memoirs of My Life*. Edited by M. M. Reese. London: Routledge and Kegan Paul, 1970.

Gitelman, Lisa. *Paper Knowledge: Toward a Media History of Documents*. Durham, NC: Duke University Press, 2014.

Godwin, William. *Diary of William Godwin*. Edited by Victoria Myers, David O'Shaughnessy, and Mark Philp. Oxford Digital Library, Oxford, 2010: http://godwindiary.bodleian.ox.ac.uk. Accessed October 1, 2017.

———. *The Enquirer*. Edinburgh: John Anderson, 1823.

———. *An Enquiry Concerning Political Justice* [1793]. Edited by Mark Philp. Oxford: Oxford University Press, 2013.

———. *Letter of Advice to a Young American on the Course of Studies It Might Be Most Advantageous for Him to Pursue*. London: Printed for M. J. Godwin, 1818.

———. *The Letters of William Godwin*. Vol. 1, 1778–1797. Edited by Pamela Clemit. Oxford: Oxford University Press, 2011.

———. *Memoirs of the Author of a Vindication of the Rights of Women* [1798]. Edited by Pamela Clemit and Gina Luria Walker. Peterborough, ON: Broadview, 2001.

Gorz, André. *Paths to Paradise: On the Liberation from Work*. Translated by Malcolm Imrie. London: Palgrave, 1985.

Grahame, James. *The Sabbath. A Poem*. Edinburgh: W. Blackwood and Longman, 1804.

Greenhouse, Carol. *A Moment's Notice: Time Politics Across Cultures*. Ithaca, NY: Cornell University Press, 1996.

Grenby, Matthew. *The Child Reader, 1700–1840*. Cambridge: Cambridge University Press, 2011.

Grenville, William. "Notes on Reading of Gillies Translation of Aristotle." BL Dropmore Papers Add. MSS 59429, unpublished manuscript.

———. "Notes on Reading of Winter 1796–7 and 1798." BL Dropmore Papers Add. MSS 59428, unpublished manuscript.

———. "Reading of Plutarch for Winter 1797." BL Dropmore Papers Add. MSS 59430, unpublished manuscript.

Griffith, Elizabeth. *The Delicate Distress* [1769]. Edited by Cynthia Booth Ricciardi and Susan Staves. Lexington: University Press of Kentucky, 1997.

——. *Essays Addressed to Young Married Women*. London: 1782.

Griffith, Richard. *The Gordian Knot*. Volume 4 of *Two Novels in Letters* by Richard Griffith and Mrs. Elizabeth Griffith. London: T. Becket and P. A. de Hondt, 1769.

——. *The Posthumous Work of a Late Celebrated Genius, Deceased*. 2 Vols. London: Printed by W. and J. Richardson, 1770.

——. *Something New*. 2 Vols. London: Printed for E. and C. Dilly, 1772.

Griffith, Elizabeth, and Richard Griffith. *A Series of Genuine Letters between Henry and Frances*. 6 Vols. 2nd ed. London: Printed for J. Bew, 1786.

The Grounds of Complaint against the Practice of Sunday Baking. London: 1794.

The Grub Street Project: Topographies of Literature and Culture in Eighteenth-Century London. Allison Muri, Director: http://grubstreetproject.net/about.php. Accessed October 1, 2017.

Guardian, The. April 18, 2007: https://www.theguardian.com/music/tomserviceblog/2007/apr/18/joshuabellnoordinarybusker.

Habermas, Jürgen. "Equal Treatment of Cultures and The Limits of Postmodern Liberalism." *The Journal of Political Philosophy* 13.1 (2005): 1–28.

——. *The Structural Transformation of the Public Sphere: An Inquiry into a Category of Bourgeois Society*. Translated by Thomas Burger. Cambridge, MA.: MIT Press, 1993.

Halsey, Katherine. "A Quaint Corner of the Reading Nation: Romantic Readerships in Rural Perthshire, 1780–1830." In *Before the Public Library: Reading, Community, and Identity in the Atlantic World, 1650–1850*. Edited by Mark Towsey and Kyle B. Roberts. Leiden, NL: Brill, 2017.

Hamersley, Rich. *Advice to Sunday Barbers against Trimming on the Lord's Day Shewing Them the Evil of That Great Sin That They Live in, by Breaking God's Holy Law, and Strict Command; By Trimming Thereon, Which Is No Work of Mercy, Nor Case of Necessity; And Therefore Ought Not to Be Done*. London: Printed Tho. Bunce, 1706.

Hansen, Mark. "Living (with) Technical Time: from Media Surrogacy to Distributed Cognition." *Theory, Culture, and Society* 26.2–3 (2009): 294–315.

Hardy, Thomas. *One Rare Fair Woman. Thomas Hardy's Letters to Florence Henniker*. Edited by Evelyn Hardy and F. B. Pinion. London: Palgrave Macmillan, 1972.

Harpold, Terry. "The Contingencies of the Hypertext Link." In *The New Media Reader*. Edited by Noah Wardrip-Fruin and Nick Montfort. Cambridge, MA: MIT Press, 2003.

Harvey, David. *The Condition of Postmodernity: An Enquiry into the Origins of Cultural Change*. Oxford: Blackwell, 1990.

Harvey, Karen. *The Little Republic: Masculinity and Domestic Authority in Eighteenth-Century Britain*. Oxford: Oxford University Press, 2013.

——. "The Manuscript History of 'Tristram Shandy.'" *Review of English Studies* 65.269 (2014): 281–301.

Haug, Frigga. "The Four-in-One-Perspective: A Manifesto for a More Just Life." *Socialism and Democracy* 23.1 (2009): 119–23.

Hawley, Judith. "Carter, Elizabeth (1717–1806)," *Oxford Dictionary of National Biography*, Oxford University Press, 2004; online edition, May 2009: http://www.oxforddnb.com/view/article/4782. Accessed February 28, 2017.

Hayley, William. *A Philosophical, Historical, and Moral Essay on Old Maids*. 3 Vols. London: T. Cadell, 1785.

Haywood, Eliza. *The History of Miss Betsy Thoughtless*. 4 Vols. London: 1751.

Henkin, David. "Hebdomadal Form: Diaries, News, and the Shape of the Modern Week." *Representations* 131.1 (2015): 52–67.

Hints for the Institution of Sunday-Schools and Parish Clubs, for the Benefit of the Poor. Chiefly Intended for the Middle Ranks Whose Benevolence May Induce Them to Attend to These Useful Institutions. London: W. Blanchard, 1789.

Hoggart, Richard. *The Uses of Literacy: Aspects of Working Class Life* [1957]. London: Penguin, 2009.

Holcroft, Thomas. *The Life of Thomas Holcroft, Written by Himself Continued to the Time of His Death from His Diary Notes and Other Parts by Hazlitt*. Edited by Elbridge Colby. London: 1825.

Howard, Carol. "'The Story of the Pineapple': Sentimental Abolitionism and Moral Motherhood in Opie's 'Adeline Mowbray.' *Studies in the Novel* 30:3 (1998): 355–76.

Inchbald, Elizabeth. *The Diaries of Elizabeth Inchbald*. 3 Vols. Edited by Ben Robertson. London: Pickering and Chatto, 2007.

———. *Memoirs of Mrs. Inchbald: Including Her Familiar Correspondence with the Most Distinguished Persons of Her Time*. 2 Vols. Edited by James Boaden. London: Richard Bentley, 1830.

———. *A Simple Story* [1791]. Introduction by Jane Spencer. Edited by J. M. S. Tompkins. Oxford: Oxford University Press, 1998.

Jenkins, Annibel. *I'll Tell You What: The Life of Elizabeth Inchbald*. Lexington: University Press of Kentucky, 2003.

Johnson, Samuel. "On the Character and Duty of an Academick." *The Johnsonians* (2000): 12–13. Reprinted from John Moir. *Hospitality: A Discourse Occasioned by Reading His Majesty's Letter*. London, 1793, 42–43.

Jones, Wendy S. *Consensual Fictions: Women, Liberalism, and the English Novel*. Toronto: University of Toronto Press, 2005.

Jupp, Peter. *Lord Grenville, 1759–1834*. Oxford: Oxford University Press, 1985.

Kelly, Gary. "Discharging Debts: The Moral Economy of Amelia Opie's Fiction." *Wordsworth Circle* 11 (1980): 198–203.

———. *The English Jacobin Novel, 1780–1805*. Oxford: Oxford University Press, 1976.

Kermode, Frank. *A Sense of an Ending: Studies in the Theories of Fiction*. Oxford: Oxford University Press, 2000.

Keymer, Tom, and Peter Sabor. *"Pamela" in the Marketplace: Literary Controversy and Print Culture in Eighteenth-Century Britain and Ireland*. Cambridge: Cambridge University Press, 2005.

Kidgell, John. *The Card*. 2 Vols. London: 1755.

Kilner, Dorothy. *Letters from a Mother to Her Children*. 2 Vols. London: John Marshall, 1780.

Kilner, Mary Ann. *A Course of Lectures for Sunday Evenings, Containing Religious Advice to Young Persons*. 4 Vols. London: John Marshall [1783?].

King, Shelly. "Portrait of a Marriage: John and Amelia Opie and the Sister Arts." *Studies in Eighteenth-Century Culture* 40 (2011): 27–62.

Kittler, Friedrich. *Gramophone, Film, Typewriter*. Translated by Geoffrey Winthrop-Young and Michael Wutz. Palo Alto: Stanford University Press, 1999.

Knight, Jeffrey Todd. *Bound to Read: Compilations, Collections, and the Making of Renaissance Literature*. Philadelphia: University of Pennsylvania Press, 2013.

———. "English Literature and the Imagined Library." Unpublished paper presented at "Literature and Contingency Workshop." Venice, 2016.

Knox, Vicesimus. *Winter Evenings: Or, Lubrications on Life and Letters*. 3 Vols. London: J. Richardson and Co., 1788.

Koselleck, Reinhart. *Futures Past: On the Semantics of Historical Time*. Translated by Keith Tribe. New York: Columbia University Press, 2005.

———. *The Practice of Conceptual History: Timing History, Spacing Concepts*. Translated by Todd Samuel Presner and Others. Palo Alto: Stanford University Press, 2002.

Krämer, Sybille. "The Cultural Techniques of Time Axis Manipulation: On Friedrich Kittler's Conception of Media." *Theory Culture Society* 23.7–8 (2006): 93–109.

Lackington, James. *The Confessions of J. Lackington*. London: Richard Edwards, 1804.

———. *Lackington's Catalogue for* 1797. London: Lackington, 1797.

———. *Memoirs of the First Forty-Five Years of the Life of James Lackington*. London, 1791.

Lamb, Charles. *The Letters of Charles and Mary Lamb*. London: Methuen, 1905.

Landon, Richard G. "Small Profits Do Great Things: James Lackington and Eighteenth-Century Bookselling." *Studies in Eighteenth-Century Culture* 5 (1976): 387–99.

Latour, Bruno. *An Inquiry into Modes of Existence*. Translated by Catherine Porter. Cambridge, MA: Harvard University Press, 2013.

———. "On Interobjectivity." *Mind, Culture, and Society* 3:4 (1996): 228–44.

Leach, Edmund. "Two Essays Concerning the Symbolic Representation of Time." In *Rethinking Anthropology*. London: Athlone Press, 1971: 124–36.

Lee, Sidney. "Griffith, Richard (*d.* 1788)." Revised by Ian Campbell Ross. *Oxford Dictionary of National Biography*. Oxford University Press, 2004: http://www.oxforddnb.com/view/article/11608, accessed 18 Jan 2017.

Lennox, Charlotte. *The Female Quixote* [1752]. Edited by Margaret Dalziel. Oxford: Oxford University Press, 1998.

Levine, Caroline. *Forms: Whole, Rhythm, Hierarchy, Network*. Princeton: Princeton University Press, 2015.

Levy, Michelle. *Family Authorship and Romantic Print Culture*. London: Palgrave Macmillan, 2008.

Literary Atlas of Europe. Directed by Barbara Piatti and Lorenz Hurni. Forschungsprojekt ETH Zurich, Institut für Kartografie und Geoinformation: http://www.literaturatlas.eu/?lang=en. Accessed October 1, 2017.

Livingston, James. *No More Work: Why Full Employment Is a Bad Idea*. Chapel Hill: University of North Carolina Press, 2016.

Looser, Devoney. *Women Writers and Old Age in Great Britain, 1750–1850*. Baltimore: Johns Hopkins University Press, 2008.

Low-Life: Or One Half of the World Knows Not How the Other Half Live, Being a

Critical Account of What is Transacted by People of All Religions, Nations, Circumstances, and Sizes of Understanding in the Twenty-Four Hours between Saturday-Night and Monday-Morning. In a True Description of a Sunday As It Is Usually Spent Within the Bills of Mortality. Calculated for the Twenty-First of June [1757?]. 3rd ed. London: Printed for John Lever, 1764.

Luhmann, Niklas. *Art as a Social System*. Translated by Eva M. Knodt. Palo Alto: Stanford University Press, 2000.

———. "The Future Cannot Begin: Temporal Structures in Modern Society." *Social Research* 43:1 (1976): 130–52.

———. *Love as Passion: The Codification of Intimacy*. Translated by Doris Jones. Cambridge, MA: Harvard University Press, 1987.

———. *Social Systems*. Translated by John Bednarz Jr., with Dirk Baekner. Palo Alto: Stanford University Press, 1995.

———. *Soziologische Aufklarung*. Vol. 6, *Die Soziologie und Der Mensch*. 2nd ed. Wiesbaden: Verlag for Sozialwissenschften, 2005.

Lurz, John. *Death of the Book: Modernist Novels and the Time of Reading*. New York: Fordham University Press, 2016.

Lynch, Deidre. *Loving Literature: A Cultural History*. Chicago: University of Chicago Press, 2014.

Major, Emma. "The Life and Works of Catherine Talbot (1721–70): 'A Public Concern.'" In *Religion and Women in Britain, 1660–1760*. Edited by Sarah Apetrei and Hannah Smith. Farnham, UK: Ashgate, 2014.

Mak, Bonnie. *How the Page Matters*. Toronto: University of Toronto Press, 2011.

Mandell, Laura. "Prayer, Feeling, Action: Anna Barbauld and the Public Worship Controversy." *Studies in Eighteenth-Century Culture* 38 (2009): 117–42.

Manovich, Lev. *The Language of New Media*. Cambridge, MA: MIT Press, 2001.

Mapping the Republic of Letters. Led by Paula Findlen, Dan Edelstein, and Nicole Coleman. Stanford Digital Repository: http://republicofletters.stanford.edu/index.html. Accessed October 1, 2017.

McCrossen, Alexis. *Holy Day, Holiday: The American Sunday*. Ithaca, NY: Cornell University Press, 2000.

McGann, Jerome. "Philology in a New Key." *Critical Inquiry* 39.2 (2013): 327–46.

———. *Radiant Textuality: Literary Studies after the World Wide Web*. New York: Palgrave Macmillan, 2001.

McGill, Meredith, and Andrew Parker. "The Future of the Literary Past." *PMLA* 125.4 (2010): 959–67.

McGurl, Mark. "Being and Time-Management." Talk given at the English Institute, September 2016.

McKenzie, D. F. *Bibliography and the Sociology of Texts*. Cambridge: Cambridge University Press, 2009.

Mee, Jon. *Conversable Worlds: Literature, Contention, and Community 1762 to 1830*. Oxford: Oxford University Press, 2011.

———. "The Press and the Danger of the Crowd: Godwin, Thelwall, and the Counter-Public Sphere." In *Godwinian Moments: From the Enlightenment to Romanticism*. Edited by Robert Maniquis and Victoria Myers. Toronto: University of Toronto Press, 2016.

Meillassoux, Quentin. *After Finitude: The Necessity of Contingency*. Translated by Ray Brassier. London: Continuum, 2008.

Mercier, Louis-Sébastien. *Memoirs of the Year Two Thousand Five Hundred*. Translated by W. Hooper. Philadelphia: Printed by Thomas Dobson, 1795.

Meyers, Victoria, David O'Shaughnessy, and Mark Philp. "Introduction." *William Godwin's Diary*. Oxford: Oxford Digital Library, 2010: http://godwindiary.bodleian.ox.ac.uk/introduction.html#Reading.

Miller, Stephen. *The Peculiar Life of Sundays*. Cambridge, MA: Harvard University Press, 2008.

Min, Eun Kyung. "Giving Promises in Inchbald's *A Simple Story*." *ELH* 77.1 (2010): 105–27.

Molesworth, Jesse. *Chance and the Eighteenth-Century Novel: Realism, Probability, Magic*. Cambridge: Cambridge University Press, 2010.

Morris, William. *News from Nowhere and Other Writings*. Edited by Clive Wilmer. London: Penguin, 1993.

Multigraph Collective, The. *Interacting with Print: Elements of Reading in an Era of Print Saturation*. Chicago: University of Chicago Press, 2018.

Muñoz, José Esteban. *Cruising Utopia: The Then and There of Queer Futurity*. New York: New York University Press, 2009.

Myers, Sylvia Harcstark. *The Bluestocking Circle: Women, Friendship, and the Life of the Mind in Eighteenth-Century England*. Oxford: Clarendon, 1990.

Nightingale, Florence. *Cassandra* [1860]. New York: Feminist Press at CUNY, 1979.

Northmore, Thomas. *Memoirs of Planetes, or a Sketch of the Laws and Manners of Makar*. London: Vaughan Griffiths, 1795.

O'Brien, Karen. *Women and Enlightenment in Eighteenth-Century Britain*. Cambridge: Cambridge University Press, 2009.

O'Connell, Lisa. "Vicars and Squires: Religion and the Rise of the English Marriage Plot." *The Eighteenth-Century: Theory and Interpretation* 52.3–4 (2011): 383–402.

Opie, Amelia. *Adeline Mowbray, or the Mother and Daughter: A Tale*. Edited by Anne McWhir. Peterborough, ON: Broadview, 2010.

Osborne, Peter. *The Politics of Time: Modernity and Avant-Garde*. London: Verso, 1995.

O'Shaughnessy, David. "*Caleb Williams* and the Philomaths Recalibrating Political Justice for the Nineteenth Century." *Nineteenth-Century Literature* 66.4 (2012): 423–48.

———. *William Godwin and the Theatre*. London: Pickering and Chatto, 2010.

Panagia, Davide. *Impressions of Hume: Cinematic Thinking and the Politics of Discontinuity*. London: Rowman and Littlefield, 2013.

Pasanek, Brad. *Metaphors of Mind: An Eighteenth-Century Dictionary*. Baltimore: Johns Hopkins University Press, 2015.

Pasenek, Brad, and Chad Wellmon. "The Enlightenment Index." In *The Eighteenth Century: Theory and Interpretation* 56.3 (Fall 2015): 357–80.

Pegge, Samuel. "The Novel's Proper Use by Young People." In *Novel Definitions, Commentary on the Novel 1688–1815*. Edited by Cheryl Nixon. Peterborough, ON: Broadview, 2009: 240–41.

Pennington, Montagu. *Memoirs of the Life of Mrs. Elizabeth Carter, with a New Edition of Her Poems . . . to Which Are Added, Some Miscellaneous Essays in Prose,*

Together with Her Notes on the Bible, and Answers to Objections Concerning the Christian Religion. London: F. C. and J. Rivington, 1807.

Perovic, Sanja. *The Calendar in Revolutionary France: Perceptions of Time in Literature, Culture, Politics*. Cambridge: Cambridge University Press, 2012.

Perry, Ruth. *Novel Relations: The Transformation of Kinship in English Literature and Culture, 1748–1818*. Cambridge: Cambridge University Press, 2004.

Peters, John Durham. *The Marvelous Clouds: Toward a Philosophy of Elemental Media*. Chicago: University of Chicago Press, 2015.

Phillips, Natalie M. *Distraction: Problems of Attention in Eighteenth-Century Literature*. Baltimore: Johns Hopkins University Press, 2016.

Philp, Mark. "Preaching to the Unconverted: Rationality and Repression in the 1790s." *Enlightenment and Dissent* 28 (2013): 1–18.

———. "Unconventional Calling, Godwin, Women and Visiting in the 1790s." In *Sociable Places*. Edited by K. Gilmartin. Cambridge: Cambridge University Press, 2017.

Piper, Andrew. *Book Was There: Reading in Electronic Times*. Chicago: University of Chicago Press, 2012.

———. *Dreaming in Books: The Making of the Bibliographic Imagination in the Romantic Age*. Chicago: University of Chicago Press, 2009.

———. "Reading's Refrain: From Bibliography to Topology." *ELH* 80.2 (2014): 373–99.

Poovey, Mary. *A History of the Modern Fact: Problems of Knowledge in the Sciences of Wealth and Society*. Chicago: University of Chicago Press, 1998.

Posnock, Ross. "I'm Not There." *Daedalus* 143.1 (2014): 85–95.

Potkay, Adam. "Determination and Agency in Fielding's *Amelia*." *The Eighteenth-Century Novel* 6–7 (2009): 335–58.

Pratt, Lloyd. *Archives of American Time: Literature and Modernity in the Nineteenth Century*. Philadelphia: University of Pennsylvania Press, 2009.

Price, Leah. "The Executor's Hand in Sir Charles Grandison." *Eighteenth-Century Fiction* 8 (1996): 331–42.

———. *How to Do Things with Books in Victorian Britain*. Princeton: Princeton University Press, 2012.

———. "Search." *Representations* 127 (2014): 93–97.

Priestley, Joseph. *Letters to a Young Man Occasioned by Mr Wakefield's Essay on Public Worship*. London: J. Johnson, 1792.

Radway, Janice. *Reading the Romance: Women, Patriarchy and Popular Literature*. Chapel Hill: University of North Carolina Press, 1984.

Rasch, William. *Niklas Luhmann's Modernity*. Palo Alto: Stanford University Press, 2000.

Rasmussen, Celia Barnes. "'Speaking on the Edge of My Tomb': The Epistolary Life and Death of Catherine Talbot." *Partial Answers: Journal of Literature and the History of Ideas* 8.2 (2010): 255–75.

Raven, James. "Arrangements for Reading and Eighteenth-Century Libraries." In *Practice and Representation of Reading in England*. Edited by Naomi Tadmor, James Raven, and Helen Small. Cambridge: Cambridge University Press, 1996: 175–201.

———. *Bookscape: Geographies of Printing and Publishing before 1800*. London: British Library, 2014.

——. *The Business of Books: Booksellers and the English Book Trade 1450–1850*. New Haven: Yale University Press, 2007.

——. "New Reading Histories, Print Culture and the Identification of Change: The Case of Eighteenth-Century England." *Social History* 23 (1998): 268–87.

——. "Selling One's Life: James Lackington, Eighteenth-Century Booksellers and the Design of Autobiography." In *Writers, Books and Trade: An Eighteenth-Century English Miscellany*. Edited by O. M. Brack Jr. New York: AMS Press, 1994: 1–23.

"Review of 'All the Works of Epictetus.'" *The Monthly Review*, November 1758, 588.

Ricciardi, Cynthia Booth, and Susan Staves. "Introduction." In Elizabeth Griffith, *The Delicate Distress*. Edited by Cynthia Booth Ricciardi and Susan Staves. Lexington: Kentucky University Press, 1997.

Richardson, Samuel. *The History of Sir Charles Grandison* [1753]. Edited by Jocelyn Harris. 3 Vols. London: Oxford University Press, 1972.

Rizzo, Betty. "'Depressa Resurgam': Elizabeth Griffith's Playwriting Career." In *Curtain Calls: British and American Women and the Theatre*. Edited by M. A. S. Schofield and C. Macheski. Athens: Ohio University Press, 1991: 120–42.

Robertson, Ben. *Elizabeth Inchbald's Reputation: A Publishing and Reception History*. London: Pickering and Chatto, 2013.

Rodgers, Katherine M. "Inhibitions on Eighteenth-Century Women Novelists: Elizabeth Inchbald and Charlotte Smith." *Eighteenth-Century Studies* 11 (1977): 63–78.

Rohrbach, Emily. *Modernity's Mists: British Romanticism and the Poetics of Anticipation*. New York: Fordham University Press, 2015.

Rose, Jonathan. *The Intellectual Life of the British Working Classes*. New Haven: Yale University Press, 2001.

Rubery, Matthew. *The Untold History of the Talking Book*. Cambridge, MA: Harvard University Press, 2016.

Sabbath Keeping: The Support of Our Religion and Nation. London: M. Folingsby, 1767.

Sachs, Jonathan. "The Glimmer of Futurity, 1811–15." In *The Regency Revisited*. Edited by Tim Fulford and Michael Sinatra. Basingstoke, UK: Palgrave Macmillan, 2016: 17–30.

——. "Scales of Time and the Anticipation of the Future." *Modern Intellectual History* 11.3 (2014): 697–718.

Sacks, Oliver. "Sabbath," *New York Times*, August 14, 2015.

Schellenberg, Betty. *Literary Coteries and the Making of Modern Print Culture, 1740–1790*. Cambridge: Cambridge University Press, 2016.

Schwarz, Tony. "Addicted to Distraction." *New York Times*, November 28, 2015.

Scott, Clive. *Literary Translation and the Rediscovery of Reading*. Cambridge: Cambridge University Press, 2013.

Sedgwick, Eve Kosofsky. "Melanie Klein and the Difference Affect Makes." In *After Sex: On Writing Since Queer Theory*. Edited by Janet Halley and Andrew Parker. Durham, NC: Duke University Press, 2011: 283–302.

Segal, Lynn. *Out of Time: The Pleasures and Perils of Aging*. London: Verso, 2013.

Serres, Michel. *The Birth of Physics*. Translated by Jack Hawkes. Manchester, UK: Clinamen Press, 2001.

———. *Hermes: Literature, Science, Philosophy.* Baltimore: Johns Hopkins University Press, 1983.

———. *Thumbelina: The Culture and Technology of Millennials.* Translated by Daniel W. Smith. Lanham, MD: Rowman and Littlefield, 2015.

Serres, Michel, with Bruno Latour. *Conversations on Science, Culture, and Time.* Translated by Roxanne Lapidus. Ann Arbor: University of Michigan Press, 1995.

Shebbeare, John. *The Marriage Act: A Novel Containing a Series of Interesting Adventures.* 2 Vols. London: J. Hodges, 1754.

Sher, Richard. *The Enlightenment and the Book: Scottish Authors and Their Publishers in Eighteenth-Century Britain and America.* Chicago: University of Chicago Press, 2006.

Sheridan, Claire. "Being Last: Widowhood and Outliving the Radical Coteries of the 1790s." *European Romantic Review* 22.2 (2011): 173–86.

Sheridan, Frances. *Memoirs of Miss Sidney Bidulph.* Edited by Patricia Koster and Jean Coates Cleary. Oxford: Oxford University Press, 1985.

Sherman, Stuart. *Telling Time: Clocks, Diaries, and English Diurnal Form, 1160–1785.* Chicago: University of Chicago Press, 1996.

Simonsen, Peter. *Wordsworth and Word-Preserving Arts: Typographic Inscription, Ekphrasis and Posterity in the Later Work.* Basingstoke: Palgrave Macmillan, 2007.

Siskin, Clifford. *The Work of Writing: Literature and Social Change in Britain, 1700–1830.* Baltimore: Johns Hopkins University Press, 1999.

Siskin, Clifford, and William Warner. *This Is Enlightenment.* Chicago: University of Chicago Press, 2010.

Six Degrees of Francis Bacon: A Seventeenth-Century Social Network. Primary Investigators: Christopher Warren, Daniel Shore, and Jesscia Otis: www.sixdegreesof francisbacon.com. Accessed October 1, 2017.

Small, Helen. *The Long Life.* Oxford: Oxford University Press, 2007.

———. *The Value of the Humanities.* Oxford: Oxford University Press, 2013.

Soni, Vivasvan. *Mourning Happiness: Narrative and the Politics of Modernity.* Ithaca, NY: Cornell University Press, 2010.

Stallybrass, Peter. "Books and Scrolls: Navigating the Bible." In *Books and Readers in Early Modern England.* Edited by Jennifer Andersen and Elizabeth Sauer. Philadelphia: University of Pennsylvania Press, 2002: 42–79.

Staves, Susan. "'Books Without Which I Cannot Write': How Did Eighteenth-Century Women Writers Get the Books They Read?" In *Women and Material Culture.* Edited by Jennie Batchelor and Cora Kaplan. London: Palgrave Macmillan, 2007: 192–213.

St Clair, William. *The Godwins and the Shelleys.* Baltimore: Johns Hopkins University Press, 1991.

———. *The Reading Nation in the Romantic Period.* Cambridge: Cambridge University Press, 2004.

Steedman, Carolyn. *An Everyday Life of the English Working Class: Work, Self and Sociability in the Early Nineteenth Century.* Cambridge: Cambridge University Press, 2013.

———. "Cries Unheard, Sights Unseen: Writing the Eighteenth-Century Metropolis." *Representations* 118.1 (Spring 2012): 28–71.

Sterne, Laurence. *The Lives and Opinions of Tristram Shandy, Gentleman* [1759–67]. Edited by Ian Campbell Ross. Oxford: Oxford University Press, 1989.

Stiegler, Bernard. "Memory." In *Critical Terms for Media Studies*. Edited by W. T. Mitchell and Mark B. N. Hansen. Chicago: University of Chicago Press, 2010: 64–87.

———. *Technics and Time*. Vol 1. Translated by Richard Beardsworth and George Collins. Palo Alto: Stanford University Press, 1998.

Stoiceff, Peter, and Andrew Taylor. "Introduction: Architectures, Ideologies and Materials of the Page." In *The Future of the Page*. Edited by Peter Stoiceff and Andrew Taylor. Toronto: University of Toronto Press, 2004.

Striphas, Ted. *The Late Age of Print: Everyday Book Culture from Consumerism to Control*. New York: Columbia University Press: 2009.

Tadmor, Naomi. "'In the Even My Wife Read to Me': Women, Reading and Household Life in the Eighteenth Century." In *The Practice and Representation of Reading in England*. Edited by James Raven, Helen Small, and Naomi Tadmor. Cambridge: Cambridge University Press, 1996:162–74.

Talbot, Catherine. *Journal*. BL Add. MS 46690, unpublished manuscript.

———. *Reflections on the Seven Days of the Week*. 2nd ed. London: J. and F. Rivington, 1770.

Taleb, Nassim. *The Black Swan: The Impact of the Highly Improbable*. New York: Random House, 2010.

Temple, William. *Diaries of William Johnston Temple, 1780–1796*. Edited by Lewis Bettany. Oxford: Clarendon, 1929.

Thompson, E. P. *Making of the English Working Class*. London: Penguin, 1979.

———. "Time, Work-Discipline, and Industrial Capitalism." *Past and Present* 38 (1967): 56–97.

———. *Warwick University Ltd.: Industry, Management and the Universities*. Edited by E. P. Thompson. London: Spokesman, 2014.

Thrift, Nigel. "Owners' Time and Own Time: The Making of a Capitalist Time Consciousness, 1300–1880." In *Space and Time in Geography: Essays Dedicated to Torston Hagerstrand*. Edited by A. Pred. Lund Studies in Geography Series B, No. 48, 1981.

Turner, Thomas. *The Diary of Thomas Turner, 1754–1765*. Edited by David Vaisey. Oxford: Oxford University Press, 1984.

Voth, Hans-Joachim. "Time Use in Eighteenth-Century London: Some Evidence from the Old Bailey." *Journal of Economic History* 57.2 (1997): 497–99.

Wajcman, Judy. *Pressed for Time: The Acceleration of Life in Digital Capitalism*. Chicago: University of Chicago Press, 2015.

Wakefield, Gilbert. *An Enquiry into the Expediency and Propriety of Public or Social Worship*. London: J. Deighton, 1791.

Walker, Eric. *Marriage, Writing, and Romanticism: Wordsworth and Austen after War*. Palo Alto: Stanford University Press, 2009.

Warner, Michael. *Publics and Counterpublics*. Cambridge, MA: MIT Press, 2002.

———. "Uncritical Reading." In *Polemic: Critical or Uncritical*. Edited by Jane Gallop. New York: Routledge, 2004.

Warner, William. *Protocols of Liberty: Communication Innovation and the American Revolution*. Chicago: University of Chicago Press, 2013.

Weeks, Kathi. *The Problem with Work: Feminism, Marxism, Antiwork Politics*. Durham, NC: Duke University Press, 2011.

Wellbery, David E. "Contingency." In *Neverending Stories: Toward a Critical Narratology*. Edited by Ingeborg Hoesterey, Ann Clark Fehn, and Maria Tatar. Princeton: Princeton University Press, 1991.

Wellmon, Chad. *Organizing Enlightenment: Information Overload and the Invention of the Modern University*. Baltimore: Johns Hopkins University Press, 2015.

Wordsworth, William. Letter to Lady Beaumont, May 21, 1807. In W. J. B. Owen, *Wordsworth's Literary Criticism*. London: Routledge, 1974.

Zerubavel, Eviatar. *The Seven Day Circle: The History and Meaning of the Week*. New York: Free Press, 1985.

Zomchick, John. *Family and the Law in Eighteenth-Century Fiction*. Cambridge: Cambridge University Press, 1993.

Zuk, Rhoda, ed. *Bluestocking Feminism: Writings of the Bluestocking Circle, 1738–1790*. Vol. 3. *Hester Chapone and Catherine Talbot*. London: Pickering and Chatto, 1999.

Index